THE
HISTORIC HOUSES
OF
PRINCE EDWARD ISLAND

SSP
PUBLICATIONS
Samson Smith

Gail Williams Cell 305-323-8991

Rm 346-8365

THE
HISTORIC HOUSES
OF
PRINCE EDWARD ISLAND

H. M. Scott Smith

Canadian Cataloguing in Publication Data

Smith. H. M. Scott, 1944-

The historic houses of Prince Edward Island

Reprint. Originally published: Erin, Ont. : Boston Mills Press, 1990

Includes bibliographical refrences and index
ISBN 0-9686803-1-3

1.   Dwellings -- Prince Edward Island -- History.   2.   Dwellings -- Prince Edward Island -- Guidebooks.   3.   Architecture, Domestic -- Prince Edward Island -- History.   4.   Architecture, Domestic -- Prince Edward Island -- Guidebooks.   5.   Historic buildings -- Prince Edward Island -- Guide-books.   I.   Title.

NA7242.P7S64 2000   728'.09717   C00-950086-3

Front cover: Arsenault House, Wellington (see p. 75). Photo: Scott Smith
Frontispiece: Photo: Lionel Stevenson
Back cover: Tyne Valley. Photo: Lionel Stevenson

Published by:
SSP Publications
Box 2472
Halifax, N.S.
B3J 3E4
Phone/Fax: (902) 429-2640

American Association
for State and Local History
Award of Merit

Winners of the
Heritage Canada
Communications Award

Designed by Gill Stead
Edited by Noel Hudson
Typography by Lexigraf, Tottenham, Ontario
Printed by Print Atlantic, Dartmouth, Nova Scotia

We wish to acknowledge the financial assistance and
encouragement of The Canada Council, the Ontario
Arts Council and the Office of the Secretary of State.

Winner of the 1992 Publishing Award, PEI Museum and Heritage Foundation

Orwell Corner.

The house is gone,
Doorstep crazy,
Boards chewed by weather.
For a while there was a hole
And then it closed
Like an eye winking
And leaving no trace.

Milton Acorn
"The House is Gone"
Dig Up My Heart, 1983

Table of Contents

Acknowledgements

Most of the research and photographs for this series of books was done in the period 1978-1981. Consequently, it was imperative to update the historical and physical status of all the buildings under our scrutiny. I would like to thank Faye Pound, Geoff Hogan and Fred Horne for their research efforts, and Archdeacon Robert C. Tuck for information and advice so graciously tendered. Gary Shutlack of the Public Archives of Nova Scotia provided information on architect Edward Elliott, and Reg Porter was most helpful in shedding some light on the career of John McLellan. I am indebted to Robert G. Hill of The Biographical Dictionary of Architects in Canada for his help and encouragement, and to the staff of the public archives of Prince Edward Island and the Prince Edward Island Museum and Heritage Foundation for their assistance.

The housing stock in Prince Edward Island has proven remarkably resilient and the pride of ownership so strong that very few houses have been lost to the ravages of time and a severe climate. As a result, most of the stunning photographs by Lionel Stevenson and Lawrence McLagan appear in this book as they were taken between 1980-1981, timeless images of a proud architectural heritage. Other photo sources are credited individually.

Many people contributed to the creation of this book. I would like to thank Reg Porter, Eileen O'Connell, Dr. Peter Ennals of Mount Allison University, Department of Geography, and Noel Hudson for their editorial help, and typists Lisa Pulsifer and Debra Murphy for somehow making it all legible. I am grateful for the drafting skills of Ross MacIntosh and Heather Mader, and the design of Gill Stead, who has once again responded in a unique and tasteful way. I would like to thank John and Jean Denison at The Boston Mills Press for their continuing faith in this project and for not compromising their production values. Thanks also to Christine Callaghan for her photographic help and unflagging enthusiasm.

Finally, I must acknowledge the financial assistance of the Canada Council and the Department of the Secretary of State for getting this project off the ground so long ago, and the Nova Scotia Department of Culture, Recreation, and Fitness for helping to keep it alive.

Dedication

This book is dedicated to the following men, whose skilled hands built many of the houses contained within these pages.

John Baker
Stephen Baker
Sylvestre Blanchard
James Campbell
Bernard Creamer
Willie James Ellis
Thomas Essery
Alex Finlay
Tom Gamble
George Gard
Harold Gay

John Goff
William Harper
James Hodgson
Robert Jones
George Judson
John Lewis
Henry Lowe
Samuel Lowe
Nathan MacFarlane
Alec "Hardwood" MacLeod
John Charles MacLeod

"Red" Hector MacMillan
Wilfred Maynard
D.R. Morrison
Richard Murley
Murdick Nicholson
Major Schurman
George Tanton
Ray Tanton
Michael Welsh
"Little Harry" Williams
Nathan Wright

Foreword

This book was born in July of 1978, when I began an investigation into the origins of Prince Edward Island architecture. Documentation of the Island's historic architecture to that point had been minimal beyond the Canadian Inventory of Historic Buildings and the efforts of a few individuals. I conducted a superficial survey of all building types in Prince Edward Island, and then journeyed to southern England and Ireland, the Highlands and Hebrides of Scotland, and the New England region of the United States in an attempt to establish the complex correlations between Prince Edward Island buildings and their antecedents. I then selected particular Island buildings for further study, based on the following parameters:

1) built before 1914;
2) a substantial and/or relevant architectural form;
3) associated with a significant historical event or personality;
4) in good physical repair and as faithful as possible to its original state.

The result will be four distinct catalogues of architecturally and historically significant pre-World War I buildings in Prince Edward Island, with brief insights into their origins. This second book follows a volume on churches, published in August, 1986. Upcoming books will focus on buildings associated with agriculture and the fishery, and on public buildings.

During the course of my investigations, I learned many things beyond the more obvious architectural conclusions. I have learned what it is that makes Prince Edward Island architecture unique: the forethought and practicality of the pioneer builders who placed decoration in a symbolic role in the context of an economical style. Lines are clean and straight, with precious little wasted space and few superfluous elements. The rolling topography and the extreme coastal climate, with heavy precipitation and high winds, have made it necessary for Island builders to look for shelter — the lee side of a hill, a protected harbour or a windbreak of trees. The idyllic postcard scene of a pastoral Island farm often does not reveal the strategy behind the building's situation.

21 Water Street, Charlotte-town,1893 when architect David Stirling was the owner. - PAPEI

21 Water Street, 1975. - PAPEI

In observing photographers Stevenson and McLagan, I have learned much about the photography of buildings and, indirectly, the visual aspects of architecture. They have taught me a different way to "see" buildings objectively, from all aspects, and that motion and haste are not conducive to a deeper appreciation of the "Queen of the Arts."

Perhaps the most significant lesson that I have learned relates to people — the hardy and resourceful generations of Scots, Irish, English and Acadians who carved communities out of the wilderness because they had a will to. This resolute adherence to the principles of practical and functional architecture, from one generation to the next, has been a profound revelation to me. These are also the people who have welcomed this inquisitive stranger into their homes.

May I apologize in advance for any errors, omissions, misinterpretations, inaccurate dates or misspellings of proper names. I regret that some communities may have been left out, but this is not a comprehensive study. I have selected only the houses that I felt could adequately illustrate this story. Of the 112 houses initially surveyed for this book, nine have either burned down (2), been demolished (2), or altered or deteriorated beyond recognition (5). This is an extremely low rate of attrition. There is no doubt that pride has maintained this housing stock, for Islanders are extremely aware of their architectural heritage. I would ask that visitors and admirers respect their property and their privacy.

Economic prosperity is well known as the enemy of architectural conservation, in that it promotes new construction. In this period of relative prosperity in Prince Edward Island, the stock of historic buildings is gradually dwindling. Although Prince Edward Island is quite advanced in the area of conservation awareness, it is most urgent that all responsible governmental heritage organizations and agencies adopt policies of inventory and education rather than legislation in order to preserve the integrity and promote sympathetic re-use of those buildings threatened with destruction or decay.

While not intended as a history text or a manual of preservation, I sincerely hope that this book and its successors will prove to be useful tools in this conservation effort. I also wish to stress that I conceived this project and these books as a beginning, in the hope that architectural historians, geographers, academics or any interested individuals might be inspired to further research and document the various aspects of architectural, social and community history touched on only briefly within the limited confines of these books. To this end, I have adopted a subjective but uncritical stance in the hope that the information contained herein will be accessible and enjoyable to everyone.

Scott Smith
Halifax, January 1991

Rogers House, Brae.

Introduction

The concept of home is a basic expression of place, individuality and of family identity. Nowhere is it more evident and real than in rural Prince Edward Island. In the lush, rolling farmland and in small villages like Tyne Valley or Belfast one can easily step back in time to the late 1700s and imagine the arrival of the Island's first settlers. With a few tools, images of their ancestral roof and dogged determination, they desperately carved primitive shelters out of the wilderness before the first snowfall of their first winter. Few of these early houses remain, but most importantly, homes had been established and roots put down. More substantial buildings took their places as each new generation spread settlement across the Island.

This book attempts to trace the evolution of that housing stock through its most humble vernacular beginnings to the high-style Victorian elegance of the urban centres. Influences affecting the form, construction and decoration of these buildings are examined in some detail, and brief biographies of the Island's most prominent pre-Modern architects are included. Although most Island houses were built of wood, brick and stone became both fashionable and practical in the latter half of the nineteenth century. A brief essay outlines this emergence. The second half of this book is a catalogue of the most historically and architecturally significant pre-1914 houses in Prince Edward Island.

The Evolution of
Domestic Architecture
in Prince Edward Island

Micmac wigwam, Rocky Point, c.1900. - PAPEI

The Micmac

The earliest form of shelter on Prince Edward Island was built by its first inhabitants, the Micmac. Their simple wigwams were generally conical in shape and circular in plan. The scale was flexible, and this type of wigwam could accommodate a single family or up to 30 people. They were built of animal hides, birchbark or reed mats, stretched over poles of spruce or pine, and were easily demountable to accommodate the migratory nature of the builders. It is interesting to note that these wigwams were assembled, erected and carried from camp to camp by the Micmac women.

A heavy leather hide served as a door at the lone entry point. A rock firepit edged with sand was built in the centre of the wigwam to provide heat in the spring and fall, but cooking usually occurred outdoors. An opening was left at the peak of the structure for smoke to escape. Several inches of fir branches overlaid by reed mats and animal furs provided a very comfortable and fragrant floor. For extra protection from the weather, these structures were sometimes lined with birchbark, a technique that was widely adopted by the European settlers who were to follow. The Island was essentially a summer fishing ground for the Micmac, so little evidence remains of any permanent camp.

Acadian Pioneers

The Acadians who arrived on Isle St. Jean in the early eighteenth century were obviously committed to a more substantial form of shelter. They were also highly skilled axemen. Timber in many species was readily available and the earliest cabins were very crude: low and tiny one-and-a-half-storey structures of horizontal or vertical logs. Corner joints were dovetailed or saddle-notched, lapped and pegged. The roof was usually constructed of logs, chinked with clay, lined with birchbark and overlain with sod or thatches of hay. The floor was hard-packed clay, and fenestration was minimal. Chimneys were commonly placed on a gable-end wall and were built of packed clay and stone rubble. Later chimneys were made of brick.

There is documentation to suggest that Jean Pierre de Roma built substantial brick dormitories at his settlement at Brudenell Point in 1732. It is also known that de Roma built houses of "piquets" or vertical logs — a house type that was probably found more commonly in other Acadian communities on the Island. Unfortunately, none of these buildings or any pre-Expulsion Acadian houses survived the destruction of 1758.

With the advent of pit sawing and crude sawmills, post-Expulsion Acadian builders, many of whom fled to the woods around Malpeque Bay to live with their Micmac friends, demonstrated considerable resourcefulness and skill. They began to build a form of half-timbered

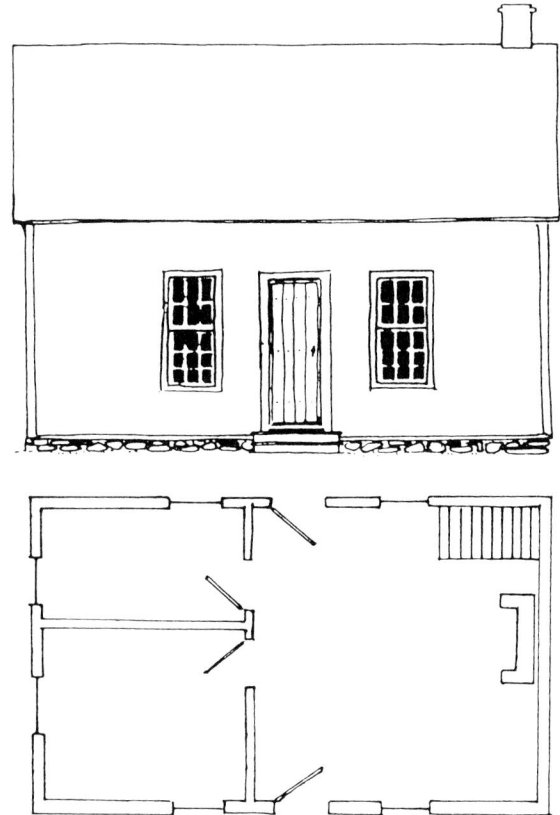

The Acadian house and plan.
- from *The Folk Legacy in Acadian Domestic Architecture* (Ennals)

stud construction called "en colombage" that had come to Acadia and New France over a century earlier. Corner joints became more sophisticated — either dovetailed or mortise-and-tenon. The builders devised a system of bridging ("entremise") between studs and filled the interstices with a mixture of clay and hay, or "wattle and daub" ("entorchis"). Many a post-Expulsion Acadian house contained a framed floor assembly.

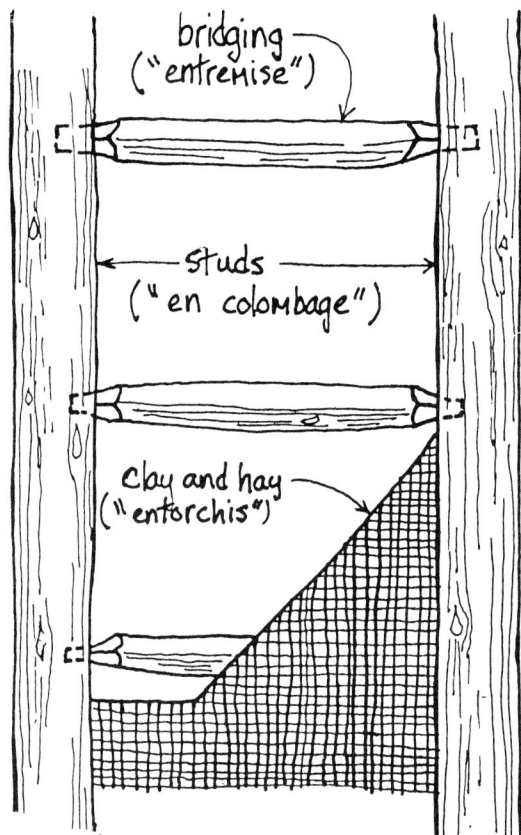

Acadian stud construction, c.1870.   - after Cunningham & Prince

Labels in figure:
- bridging ("entretoise")
- studs ("en colombage")
- clay and hay ("entorchis")

around the kitchen hearth, and when kitchen wings were later added to the basic Acadian house, family life followed, creating a more formal salon in the main house. Traditionally a room where a family would receive a visiting priest, the salon is used today for receptions and other ceremonial or festive functions. (see pp. 98, 99)

Early in the nineteenth century, Acadians began to adopt New England building techniques in their house construction. By 1850 shingles and clapboard were ubiquitous over Western and balloon framing and Loyalist dormer designs and decorative features had become quite common. The arrival of small heating stoves in the 1860s, and later central heating systems, signalled the beginning of modern architecture for Acadian builders. Family life no longer had to revolve around the hearth, effectively allowing the house plan to develop a wider range of uses, including the second floor. It was not until after World War II, however, that the renowned Acadian propensity for colour became evident.

There is little evidence that "pièces sur pièces" houses, squared horizontal logs tenoned into grooved posts, so common in the St. Lawrence Valley of New France and on Isle Royale (Cape Breton), were built at all in Prince Edward Island, except for the well-to-do or the military. Acadian houses built of stone were non-existent, as Acadian axemen lacked the skills and tools to quarry the soft Island sandstone.

A reconstruction of a typical Acadian village in the 1820s can be found at Mount Carmel in Prince County.

Generally, the Acadian farmer's family was small and so was his house, built to conserve energy. The floor plan was quite basic, a large communal multi-purpose area ("salle commune") with perhaps one sleeping chamber and a storage area partitioned off. Thus, the Acadian house was quite different from that of the Loyalist or British settler who favoured the centre-hall plan. Family life was centred

Selkirk Settlement hut reconstruction, relocated at Island Market Village, North River.
- Scott Smith

British Settlers

Immigrants from Britain arrived in waves in Prince Edward Island after the Treaty of Paris was signed in 1763. They came from the Highlands, Lowlands and the Hebrides of Scotland, the southern counties of Ireland, and Cornwall and Devon in England. They arrived aboard ships called the *Polly,* the *Lovely Nelly* and the *Annabella* with the clothes on their backs, a few tools and memories of their ancestral roof still fresh in their minds. They were soon to discover that the immediate task was to translate the stone construction of their homeland into wood. Their first efforts were quite modest and bore a striking resemblance in scale and style to those of their Acadian predecessors.

Alex Shaw, who settled at West River in the early 1800s, left this account of early cabin construction: "After having cut a small piece out of the forest as close to the river as possible, a house is erected mostly of round logs dovetailed at the corners, the chinks between the logs being tightly caulked with moss; the roof is covered with bark taken from fir or spruce trees, or else sedge grass from the marshes. The floor is the ground, smoothly packed, and a huge fireplace in the centre of the floor, with a hole in the roof for the exit of smoke."

Thatched cottage, Annestown, County Waterford, Ireland.
- Scott Smith

Staffin house, Isle of Skye.
- Scott Smith

Gradually, the stylistic models of the Old Country became more actualized, familiar elements such as dormers, root cellars, framed doors and windows, and clay brick chimneys and ovens slowly began to appear. The relative proportions of the entire building became more generous, comparable to those of their homeland. The floor plan, usually 14' x 20', was based on descendants of either the one-and-a-half-storey hall-and-parlour plan of the Northern blackhouse or the later centre-hall floor plan of the English or Irish crofter's cottage. The translation of these stone British stereotypes into wood-frame construction led the builder in a new direction, at times forcing him to improvise. This, in effect, contributed to a new regional vernacular architecture.

After 1800, settlers' houses became more substantial. Logs were squared and clad with clapboard or shingles, and shingled roofs began to appear. More solid foundations and root cellars of sandstone soon replaced timber piles. Windows were few and very small, and the only ornament to appear on the cottage exterior was rough, hand-hewn window and door trim. The British were also expert axemen. With a froe (frow) and mallet they hand-split shingles and clapboards and used their axe to shape rough furniture. Through the mid-nineteenth century, timber construction graduated from the hand-hewn stage to pit sawing and finally milled lumber.

The interiors remained Spartan but were more comfortable and commodious. Perhaps

17

MacIsaac House, Herman-ville. Four generations of MacIsaacs have successively occupied this Maritime Vernacular house since John MacIsaac built it between 1816 and 1819.

with the aid of the remaining Acadians, the British housebuilder discovered ways to insulate and line his walls with seaweed, sawdust, clay or birchbark and sheathe the interior with wide pine or spruce boards. Family life was centred around the hearth or stove. The location of the brick or stone fireplace varied from a central location to within longitudinal transverse walls to a location in the gable-end walls, depending on the particular house type. As families grew, the floor plan was subdivided into smaller rooms, or additions were built on the side or rear of the house. Families often slept on an upper landing, huddled around the chimney flue. By the 1840s, lath and plaster finishes became more common, often with a coat of whitewash. Decoration consisted of a mantle over the hearth, a crucifix, and crude baseboards or mouldings.

There are some sandstone houses on Prince Edward Island which are exact replicas of their corresponding British models, such as the Aitken house in Lower Montague (see p. 132). Rarely was a pioneer builder able to achieve this degree of replication in wood. Today we can still find log framework hidden beneath the more contemporary cladding of some of the Island's historic houses. Reconstructions of some early log buildings can be found at the Island Market Village in North River, Queen's County.

Gambrel house, Brackley Beach. - Scott Smith

The Loyalists

No single group of immigrants had more influence on the future of domestic architecture in Prince Edward Island than the Loyalists. They fled the thirteen American colonies in the period of 1783-1790 to remain loyal to the British Crown. Many came from Rhode Island, and some were enticed to relocate from the Shelburne, Nova Scotia, settlement. They settled chiefly on the Malpeque - Bedeque isthmus.

The Loyalists brought with them a distinct architecture, also deeply rooted in their British ancestry. The carpentry skills evident in their buildings and a full repertoire of neo-Classic detailing were indeed representative of the eighteenth-century New England building tra-dition. These four distinct Loyalist building types, adapted to local conditions, were to form the basis of a new regional vernacular architecture and collectively they generated a substantial influence on the early architecture of Prince Edward Island.

The saltbox is extremely rare on the Island, and the gambrel became more popular in the latter half of the nineteenth century, due at least in part to the influence of sea captains from Newport or Portsmouth, Massachusetts, visiting Island ports. By far the most common Loyalist housing form was the Cape Cod cottage and its variants. A one-and-a-half-storey frame building with a steeply pitched roof and low

Pit-sawing, an early energy crisis.
- after Cunningham & Prince.

Taper House, Georgetown. This building type is very rare on Prince Edward Island, with only two other surviving examples. Called a "saltbox" because its asymmetrical profile bears some resemblance to an old kitchen spice container, this style was quite common in eighteenth-century American Colonial architecture in New England. Sometimes referred to as a "cat's slide," the style easily accommodated an expansion to the rear, and with the long roof slope oriented to the north, it provided extra environmental protection as well.
- sketch by Robert C. Tuck

Above left: Standard variations of the Cape Cod House as revealed in the south elevation; (left to right) house, house-and-a-half, double-house. Above right: Standard variations of the gable observed on the Lower Cape. The two at the left are characteristic of the house and house-and-a-half; the two at the right of the double house.
- Ernest Allan Connally

Brien House, Victoria, built c.1875, illustrates the rigid symmetry and proportional refinement of a Colonial Georgian house.

walls, the Cape Cod had little or no overhang at the gable end or the eaves (see p. 104). Fully developed, the Cape Cod was usually a five-bay structure with a centred entrance and chimney stack, commonly 36′ x 24′. Its lesser variants were the "three quarter house," with two bays on one side of the entrance and one on the other, and the "half-house," with two windows to one side of the entrance. The Cape Cod form was also extremely accommodating in terms of adding extra bays or dormers, consequently many hybrid Cape Cod houses can be found, if not readily identified, on the Island.

The Colonial Georgian house was introduced by the Loyalists, but its image was reinforced considerably, particularly in urban situations, by continued immigration from Britain in the first half of the nineteenth century. A two-and-a-half-storey, five-bay building with a centre-hall floor plan, the Colonial Georgian house was proportionately refined and rigidly symmetrical. Steep-gabled roofs were more common than hipped roofs, and the detailing, while somewhat more restrained than that of the formal British tradition or even its American predecessors, was decidedly neo-Classic. (see also p. 92)

House on Second Street, Summerside.   - sketch by Robert C. Tuck

Maritime Vernacular

The birth of a distinct regional architecture was therefore a metamorphosis of two separate but not dissimilar building traditions, the Cape Cod cottage or Colonial Georgian house of the Loyalists and the post-1800 stereotypes of the British settlers. The Maritime Vernacular house was the immigrant builders' direct response to local materials and conditions and to other housing forms that they observed. The one-and-a-half-storey, three-bay building had a shallower pitched roof and higher walls than the Cape Cod and its plan was based on the symmetrical centre-hall plan of the Colonial Georgian house. A variety of dormer styles provided additional second-floor space, usually for a bathroom. It is difficult to affix a chronological period to the establishment of the Maritime Vernacular house, but surviving examples seem to indicate that in Prince Edward Island it was built in the period 1810-1880.

Belmont, East Royalty.   - Lawrence McLagan

The British Classical Tradition

With an overwhelming increase in immigration from Britain in the first half of the nineteenth century, the American influence on domestic architecture on the Island gradually diminished. Concurrently, the expression of stability and permanence became the prime objectives of British government officials, architects and builders in Atlantic Canada. Inspired by the work of British architects James Gibbs, John Nash, Robert Adam and John Soane, regional architects promoted Palladianism and Greek and Roman Classicism in the construction of their public and commercial buildings. Province House in Charlottetown, built between

Woodbine Cottage, Port Hill.

- from *Meacham's Atlas*

1843 and 1848 and designed by architect Isaac Smith, is a good example. In domestic architecture, beyond the pure Georgian house of the Loyalists, the Classic Revival was expressed through the application of a vast repertoire of neo-Classic details: pediments, porticos, fan lights, corner pilasters, eave returns, Palladian (or Venetian) windows, architraves, and the implication of the classical orders in wooden columns or pilasters. This vocabulary of elements was applied to not only the prototypical two-and-a-half-storey, gabled-roof, symmetrical building form, but to vernacular houses as well. The popularity of the Classic Revival idiom spread far beyond the exclusive means of the urban wealthy.

The Classic Revival also helped to bridge a stylistic gap between the monumental formality of the Georgian style and the romanticism of the Victorian period. In fact, some historians regard the Greek Revival style and the "temple house" as a precursor of the Romantic movement (see above and p.78).

24

Watermere, Charlottetown, designed by architect William Harris and built in 1877.
- Lawrence McLagan

Victorian Architecture

Inspired by the work of the English architect Pugin and the writings of Ruskin and A.J. Downing, North American architects sought to explore new frontiers in design. They were also heavily influenced by the Ecclesiologists, who favoured the "True Pointed and Christian style" in church design. The latter half of the nineteenth century was a period of prodigious economic growth, and architects and builders were eager and extreme in their rejection of the symmetry and ordered elements of neo-Classicism. Irregularity, emphasis on verticality and a picturesque taste became the norm. The onset of the Victorian period (1837-1901) heralded a succession of increasingly elaborate architectural styles that swept western Europe and North America. The first such style was the Gothic Revival, beginning about 1845. Drawing on its religious medieval roots, the Gothic Revival embodied an extensive vocabulary of ornamental bargeboards, pointed arches, steeply pitched gables and dormers, Gothic windows, finials and lacy gingerbread trim. Gothic Revival houses are quite commonplace on the Island, particularly the "ell" farmhouse and the Picturesque cottage. The efficient asymmetrical plan of the former was extremely receptive to additions and thus became a prin-

Left: Brown House, York. This fine Gothic Revival-style house was built c.1870 by David Brown. Right: Richmond double ell, Mom's Bed and Breakfast.   - Scott Smith

ciple design employed by growing farm families. In fact, many buildings became ell farmhouses with the addition of a secondary wing, usually housing a kitchen. Seldom, however, was the main house connected to a barn or outbuildings, a situation commonly found in New England. Ornamentation was usually Gothic, although some intriguing eclectic mixes can be found, creating a peculiar vernacular interpretation of an essentially Gothic Revival form.

The Victorian period not only promoted innovation, imagination and experimentation in design but it also nourished, through a wider vocabulary of elements and a variety of form, colour and texture, more individualistic expression. The Picturesque movement, born in Britain in the late eighteenth century, only began to affect regional domestic architecture toward the middle of the nineteenth century. It was really an attitude toward architecture promoted

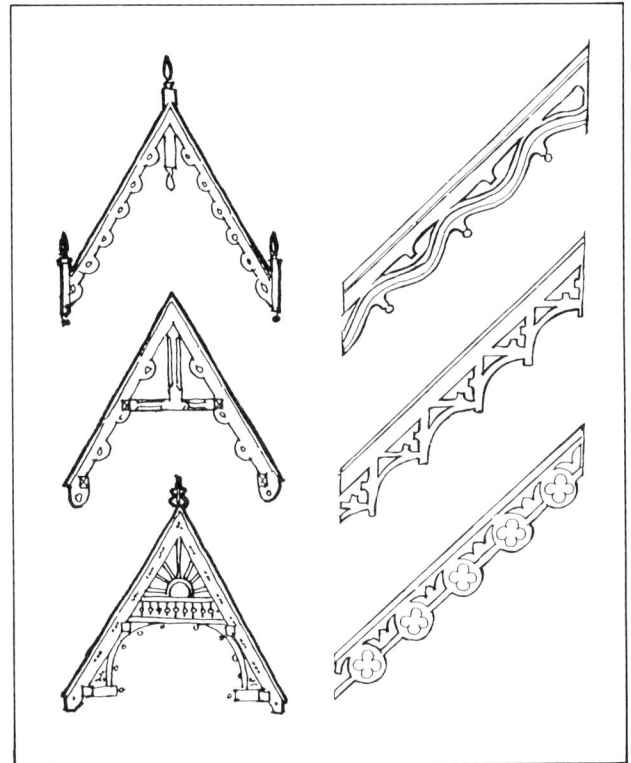

Bargeboard Gothic trim and gingerbread.   - Scott Smith

Detail, Tuplin-Lefurgey House, Summerside.   - Lawrence McLagan

Fairholm, Charlottetown.   - from *Meacham's Atlas*

by artists and philosophers in which the building itself became subservient to its own status in the landscape. Thus the building was to become, ideally, more natural, honest and painterly. The "cottage ornée" was basically a suburban form favoured by the middle and upper class. In Atlantic Canada and on Prince Edward Island, this ideal was rarely found and only approximated. Most commonly it was represented by a board-and-batten Carpenter-Gothic cottage or a neo-Classic one-and-a-half-storey cottage with a Gothic centre gable that fit naturally into the accommodating and stunning Island landscape. The Picturesque embraced any style, eclectic or not. Taste gradually turned from the symmetry of the Gothic cottage to the asymmetry of the Italian villa. Although Georgian in form, "Fairholm" in Charlottetown (see above) perhaps most closely approximates the ideal Picturesque villa.

The Italianate period reached its zenith in Canada around the time of Confederation. Based squarely on the architecture of the small villas of Tuscany in northern Italy, the Italianate was extremely popular not only in commercial and public buildings but in the pattern books for house design. Characterized by single or

27

Doyle House, Georgetown. This two-and-a-half-storey Italianate house was built in Georgetown in 1906 by Dr. Duncan Stewart.

paired wide eave brackets and tall, slender, round-headed windows, shallow-pitched hipped roof, a balcony, verandahs and a tall, square watchtower or belvedere, the Italianate house was not necessarily symmetrical in plan. Today in Prince Edward Island, a pure Italianate house is quite rare. More commonly it is found as an altered style, combining elements of the Second Empire style, with which it gradually merged, and those of the Queen Anne Revival.

The double-pitched mansard roof, designed to create more space in second-floor rooms, is the most prominent feature of the brief (1860-1880) but flamboyant Second Empire style. This was an international style that originated in the Second Empire of Napoleon III in France in the 1850s and quickly spread to England and North America. Initially promoted for use in commercial and public buildings, the Second Empire was quickly embraced by architects and builders as the fashion for high-style residences.

Irving House, Alberton, built by John Donald, c.1875. - Scott Smith

MacLeod House, Valleyfield. This delicately proportioned Second Empire-style house was built c.1910 by Alec "Hardwood" MacLeod.
- Scott Smith

A Second Empire house presented a high-profile, decorated image but also one of stability and mass. Tall central towers and frontal pavilions with iron roof cresting were quite common. In many cases the lower roof slope of the mansard was bellcast or concave. (see p. 161)

In the Atlantic region, the Second Empire house took on a lighter, more whimsical presence. With the proliferation of wood construction it proved quite accommodating to vernacular interpretation and more elaborate detailing. Regional characteristics quite common on the Island are the two-storey, three-sided bay projection and the deep-set dormer window that cut the eave line.

The Second Empire style abruptly fell out of fashion. Perhaps it was the aesthetic restlessness of the Victorian period that could not long tolerate its dual identity of stability and flamboyance.

Growing out of a marriage between the Queen Anne style of English vernacular domestic architecture and American Colonial Revivalism, the Queen Anne Revival in Canada epitomized the excessiveness of the Victorian period. A riotous amalgam of irregular massing, steep hipped roofs, gables, pediments, offset towers and broad verandahs, the Queen Anne Revival presented a very complex and eclectic visual appearance. External surface textures and patterns varied greatly, sometimes mixing brick, shingles and clapboard on the same facade. A variety of window types, including the Palladian, could be found on a Queen Anne house.

The Queen Anne Revival and its variants, the Stick style and the Shingle style, were commonly found in architectural pattern books until the turn of the century. Their usage was not confined to urban situations, and with the

Judson House, Alexandra. (see also page 138)

exception of the Shingle style, several good examples can be found in rural Prince Edward Island. Characterized by external structural expression in wood, half-timbered detailing on the facades, and angular two-storey bay projections, the Stick style was promoted by American architect Henry Hobson Richardson in the 1870s. The Shingle style, also a Richardsonian-inspired movement, appeared well into the twentieth century in Canada. It simplified the rambling Queen Anne style and was welcomed as somewhat of a relief. Its complex, multi-gabled roofline, shingle cladding on walls and roof, and stone foundations were not often found on Prince Edward Island, although the McNichol/Best house in Cardigan is a good example. (see p. 128) The Shingle style was used primarily in the design of large summer houses.

Shortly after the turn of the century, architects and builders became less and less tolerant of the excesses of the Victorian period. Public taste had changed as well, and reaction was swift. A simplification of form became almost immediately evident. In the Maritimes, the Four

Rogers -MacDonald House, Alberton (now Buchanan Lodge).

Carruthers House, Hamilton.

Square style emerged as a welcome relief to the Queen Anne of the High Victorian period. A two-and-a-half-storey building that was square in plan, the Four Square was characterized by a steep hipped roof, a large front verandah over an eccentric entrance, and large dormers with hipped roofs. The Four Square is quite common in Prince Edward Island, par-

ticularly in the western half of the province. Its form found favour with the new breed of silver fox farmer, whose sudden wealth initiated by Robert Oulton and Charles Dalton in the early 1900s, allowed them to build such large, rambling houses. The Island "fox house" was essentially a Four Square building with superimposed Colonial Revival details. Second-storey balconies surmounted by neo-Classical pediments were often later additions. The good fortune gained by a select few in the period 1905-1930 created an architectural wave that spread quickly across western Prince Edward Island. These sprawling twelve- or fourteen-room houses were often elaborately appointed. In some cases, "fox house" details were imposed on buildings of vastly different styles with predictable results. They seemed to suggest that the wealthy fox farmers of Tignish, Alberton, Montrose and Summerside were not at all shy about architectural experiment or juxtaposition.

Above: Cameron House, Canoe Cove, formerly St. Thomas Anglican Church, Long Creek. Below: MacLennan House, Wood Islands.
                                                    - Scott Smith

Gaudet House, Miminegash (now demolished).

Originating in Holland, transposed to the East Anglia region of England in the seventeenth century and brought to New England by religious dissenters, the gambrel house (see above) was a Loyalist import that was not particularly common in Prince Edward Island. At least one gambrel design could be found in every Victorian pattern book, however, and there are a few good examples of this turn-of-the-century house type extant in the eastern part of the Island. These buildings are commonly the ell farmhouse type with additional second-storey space provided by the expanded roofline.

Prince Edward Island is rich in vernacular, neo-Classical and Victorian domestic architecture, but there are also some anomalies, oddities and hybrid forms in their midst. There are many examples of the free-spirited, imaginative style of some of the Island's more independent builders — those who saw a particular solution, usually based on practicality rather than whimsy, to a particular problem. The half-scale fisherman's house, (see above) for example, could almost be classified as a Maritime vernacular form. The

Simmons House, Wilmot.

octagonal house at New Perth (see p. 135), the angular Shaw house in Souris (see p. 153) and the Cameron church/cottage at Canoe Cove (see p. 32) are typical of the ingenious solutions to practical building problems. There are countless examples of barn-to-house conversions and other cases of churches, railroad stations and even old schoolhouses being transformed into private dwellings. There are indigenous interpretations of period styles, such as the Orr/Deblois house in Dalvay (see p. 99) and the Lavandier house in Georgetown (see p. 148). Styles collide, sometimes successfully, in *hybrid* houses such as the Bourke house in Millview (see p. 137) or the MacLennan/Hunt house in Summerside (see p. 84). There are *composite* houses, such as the Simmons house in Wilmot (see above) or the Clow house in Murray Harbour North (see p. 142) where buildings of two distinct styles are joined together. Others, such as the Affleck house in Lower Bedeque (see p. 89), have had elements of period styles added to a more modest vernacular shell.

MacCallum House, Brackley Beach.

Stone and Brick

Despite the fact that beds of red sandstone of the Upper Carboniferous Period underlie practically all of Prince Edward Island, there are remarkably few examples of buildings constructed of this material. In the early to mid-nineteenth century on the Island, there was a very brief movement to build farmhouses out of sandstone that had been locally quarried and dressed. Prior to this period, the use of sandstone as a building material on Prince Edward Island was confined almost exclusively to the construction of foundations for barns and farmhouses and for culverts, bridges, fencing material, well linings, and decorative uses. The stones were usually quarried on a particular farm property, at a very shallow depth, and hauled to the site on a rock sleigh drawn by horses.

These houses were built usually by itinerant or imported British masons for the military or English, Irish or Scottish owners of some financial means. For the most part, their design was based upon a variation of the one-and-a-half-storey, gabled-roof farmhouse built in Britain

Rose Valley house, Dunvegan, Isle of Skye, Scotland. - Scott Smith

Young House, Charlottetown, designed by W.C. Harris and built in 1887. - Scott Smith

in the eighteenth and nineteenth centuries and almost without exception, they were of the centre-hall plan. The location of fireplaces and their flues varied from within the gable-end walls to a central location, but some were built into interior partitions. The house dimensions ranged between extremes of 40' x 30' and 30' x 20'. They were expertly and sturdily built with walls two to three feet thick, most commonly in vermiculated coursing. The masons showed great skill in quarrying the stone and dressing it with such precision. Their task was simplified, no doubt, by the softness and porous quality of the stone, and today one can still see chisel marks on many of the stones. Creative masons have, at times, attempted some low relief sculpting, particularly on the front elevation. This highly textured surface, combined with the rust-red of the stone, the white mortar and the occasional green or yellow lichen or rock moss, renders a vivid and highly satisfying visual impact (see p. 58).

Apart from churches and public buildings, we have been left with a small and endangered legacy of twelve small Island houses built in whole or in part with Island sandstone. Three others have been demolished within the last 50 years, including a large ten-room, two-and-a-half-storey Georgian estate in Malpeque called "Stewart Hall," which was built in 1820. Two of the houses are unoccupied; the Profitt house in Burlington is in a forlorn and derelict state, and the Parker house in Georgetown is suffering from its neglect. The Pickering house in Clinton and "Newstead" (now the MacPherson residence) in Winsloe are two houses that still embody stone wall sections of the original buildings. The Young house in Charlottetown, built in 1887 and designed by W.C. Harris, is constructed for the most part of Wallace, Nova Scotia, freestone, but the windows and doors are framed in Island sandstone.

Half-round window, Barton Lodge, South Winsloe.

Beneath a coat of brownish-red stucco, the Haslam house in Springfield is a typical Island stone house. It was built in 1856 by Robert Haslam, the sixth son of Thomas Haslam, a native of Queen's County, Ireland, who settled on Prince Edward Island in 1818 and gave Springfield its name in 1828. The house was purchased in 1956 by the Sinclair family, whose ancestors immigrated to the area from Scotland in 1840.

The MacCallum house in Brackley Beach was built before the middle of the nineteenth century by two itinerant Scottish masons for Captain James MacCallum, whose father, Duncan, immigrated to Prince Edward Island from Scotland in 1770. "Linden Cottage" underwent a rather drastic renovation in the early 1950s, at which time a kitchen wing was demolished and the rear shed dormer added. The three frontal shed dormers were built in the 1930s. There is a feeling of stability and permanence about this house. The stones have been cut with precision and laid with great skill and care, because the walls are still straight and true. It stands in a lovely pastoral setting surrounded by linden trees imported from Scotland. (p. 34, 113, 164)

Barton Lodge, located in South Winsloe, was built by an Englishman, William Buxton, c.1846. A one-time roadhouse, it shows evidence that a rear six-room kitchen wing and front verandah were demolished at the turn of the century. The house apparently was designed by Isaac Smith, one of the architects of Government House in Charlottetown, and its gable-end walls contain some attractive half-round or crescent windows (above).

Front door, Bagnall House, Hazelgrove.

The Bagnall house in Hazelgrove sits at the top of a rise in a grove of trees just outside Hunter River in central Queen's County. It was built in 1851 by Edwin Bagnall, a descendant of United Empire Loyalists. An interesting feature is the narrow front door with fan light and inscribed stone lintel above. The Aitken House at Lower Montague, the Atwell House in Clyde River, and the Ramsay House on the Beech Point Road, near Hamilton, are featured on pages 132, 110, 87 respectively, in the catalogue section later in this book.

As ownership changed or as families grew, additions were made to the original houses, usually in wood-frame construction. Probably the most popular of these was the addition of a central dormer to the roofline, undoubtedly to generate more living space and light on the second floor. Renovations usually included the conversion of bedroom to bathroom and a kitchen enlargement, as the original kitchens tended to be quite small. Because of the solidity of its exterior walls, this house type was not easily extended. There is evidence that most of these houses had a rear kitchen wing at one time, but many have been removed. Some front verandahs have also disappeared, and dormers have had to be rebuilt. Obviously, the marriage between stone and wood in these cases was not a happy one. Moisture infiltration and drafts around door and window frames and under the gable-end eaves are further evidence that builders had some difficulty with the structural joints between wood and stone.

The main problems in local masonry construction appeared to be the settling of foundations in the soft Island soil, and the lack of waterproofing or a perimeter drainage system around the outside foundation walls. There has been a common disintegration of the gable-end walls and this may be attributable to the settling of foundations, the loading and thermal effect of chimneys, or the omission of structural ties between the wooden floor assemblies and the outside walls. In houses where these problems have been rectified, stone masonry construction in Prince Edward Island has proven to be very efficient thermally: cool in the summer and warm in the winter. Houses with perimeter cavity walls (one-inch air space or more) have been much drier than their counterparts with solid stone walls. One of the most positive aspects of these stone houses has been their compact and efficient floor plan. The centre-hall plan has been time tested, and its symmetrical form simplified construction considerably. As well, it has proven to be heated easily and efficiently from a central source. For an Island family of a particular and fixed size, it seemed ideal.

In the latter half of the nineteenth century and the early part of the twentieth, there arose considerable discussion and controversy regarding the use of Island sandstone as a building material. The Great Fire of 1866 in Charlottetown, which destroyed four blocks of wooden frame buildings, and subsequent fires in 1884 and 1887, caused city administrators to re-examine not only existing firefighting practices but also the methods and materials of construction. Stone and brick in particular were looked upon more favourably. Summerside also experienced a severe fire, in 1906, when it lost 155 buildings. A number of rural church devastations prompted Island architect W.C. Harris to recommend a more widespread use of Island sandstone in the construction of his churches. In a letter to the *Daily Examiner* in 1888 he referred to St. Paul's church in Sturgeon, built the previous summer of Island sandstone: "in the severe weather of last winter, [the walls] were coated two inches in thickness with fro-zen sleet for two or three months [a severe test for stone walls] and the stone stood the test without the slightest damage." Of his inspection of a stone house "on a farm opposite to Georgetown," probably the Aitken house, he wrote: "The stones in the walls were perfect, still retaining the tool marks made when they were cut half a century ago. The inmates said that the building was always comfortably warm in winter and never damp."

In spite of its fine aesthetic qualities and because of increased competition from Island-made bricks as a fire-stopping building material, stone construction had become impractical and uneconomical. In 1834 there were only two brick buildings in Charlottetown. In 1861 there were nine brick kilns in operation in the province, producing 1.3 million bricks. Prior to the fiery holocausts in Charlottetown and Summerside, brick was used primarily in the construction of chimneys and foundations and by the more progressive merchants in Charlottetown. But by 1878 there were seven kilns operating in Southport alone. After the fires in Charlottetown in the 1880s, insurance companies, civic officials and local newspapers exerted great pressure on merchants to build in brick. It remained nevertheless an urban building material, favoured primarily by merchants and the well-to-do. Consequently, rural houses built of brick were, and still are, quite rare. Good brick clay was in short supply, except in the swamps and barrens of Prince County. Several brick buildings disintegrated and were either demolished or rebuilt. The most durable brick, on the evidence, was of the imported variety, but the extra costs associated

Richard House, Tignish. Built concurrently with St. Simon and St. Jude Roman Catholic Church, c.1860, this small centre gable cottage was also built with brick from the local brickyard belonging to Francis Hughes.

- Scott Smith

with this importation discouraged many a builder. Wood-frame construction remained the most popular and economical mode of construction.

A report on the ornamental stones of Canada, by William A. Parks in 1914, did not enhance the reputation of Island sandstone. His conclusions were based on the test results of one sample taken from Henry Swan's quarry, just outside Charlottetown. He stated: "The sandstones are, for the most part, of coarse grain and inferior durability. Further, they frequently occur in thin layers interstratified with shales, which materially increases the cost of extraction." The report, most decidedly negative, continued, "the crushing strength is very low and the loss of strength on soaking is remarkable." Mr. Parks, however, concluded

with some encouraging words: "The stone is easily worked by picks and wedges and it is quarried almost entirely without the use of explosives . . . the lower stone is said to be harder and more durable."

Although most of the Island's rural quarries are now overgrown and are difficult to find, there is still a glimmer of hope. If the stone can be quarried economically from deeper strata, and skilled stonecutters and masons can be found locally, the author believes that Island sandstone can overcome its somewhat tainted reputation and that a revival in stone construction is a distinct possibility. It is hoped that a practical and economical solution may be found so that we may once again see handsome rust-red buildings rising out of the Island's red clay soil. (see pp. 56, 60)

A.J. Downing design.
- from *Cottage Residences*

Influences

We have seen the significant impact that British immigration and the Loyalists have had on domestic architecture in Prince Edward Island. We are also aware of the widespread proliferation of universal styles in the Victorian period, particularly in Charlottetown and Summerside. But what were some of the other forces and factors at work that helped to shape the styles and forms of our architectural heritage? Unquestionably, the most significant influence was the printed word. Architectural books such as Minard Lafever's *Modern Builder's Guide* (1833), John Plaw's *Ferme Ornée* (c.1800), Calvert Vaux's *Villas and Cottages* (1857), and Samuel Sloan's *The Model Architect* (1853) all had a great impact on architect and builder alike. The most influential of all these writers, however, was American landscape gardener and "architectural composer" A.J. Downing. His

Cottage Residences and *Architecture of Country Houses*, written in the 1840s, were landmark publications. Downing and his collaborator, architect A.J. Davis, were instrumental in establishing an attitude of expressing truth, honesty and the picturesque in domestic architecture. They were fountainheads in the instigation of the revivalist spirit, both Gothic and Italian.

Agricultural journals and magazines, such as *The Agriculturalist*, *The Cultivator* and *The Canadian Farmer* (1864), began to reach the Island in the latter half of the nineteenth century, and many of these contained sections or some other reference to architecture or farmhouse construction. By the 1870s Victorian building catalogues, manuals and pattern books were readily available. The most popular of these were written by builders such as Bicknell and Comstock (1873 and 1881), Ogilvie (1885) and the Palliser Brothers (1876-1908). Hodgson's *Encyclopedia Of Carpentry And Building* (1903) was also widely used.

The previously noted fox-farming boom of the early 1900s brought considerable wealth to western Prince Edward Island. In their business trips to London, Paris and New York, Island fox ranchers were exposed to the High Victorian architectural trends that were sweeping the western world.

The growth of the shipping and shipbuilding industries in the 1860s saw not only American and British sea captains sail into Island ports, but highly skilled ship's carpenters and joiners as well. They undoubtedly brought with them new and unique concepts of construction and design, but also a significant stimulus to the economies of such communities as Port Hill, Summerside, Charlottetown, Victoria, Georgetown and Souris.

Finally, there was the resourcefulness, skill and imagination of such master builders as Robert Jones, Wilfred Maynard, Harry Williams, George Gard and Thomas Essery that made some Island houses unique. These men had a way with wood — techniques that could not only interpret styles imposed by others but also render them in a way that was lasting and special.

Eleven Architects

Thomas Alley

Thomas Alley was born in Charlottetown in 1820 and had a long career as a builder-architect. He was Superintendent of Works for the province in 1866. His most noteworthy buildings are in Charlottetown: the Union Bank, his own residence at 62 Prince Street, Trinity Church, and the former Law Courts Building (1874-1876), now the Hon. George Coles Building. There is evidence to suggest that he designed St. Patrick's rectory at Fort Augustus (see p. 157), which bears a striking resemblance to his Prince Street residence. An avid reader, he became blind late in life and died in 1900. There is a street named after him in Charlottetown.

George Baker

George E. Baker was born in Bedeque, Prince Edward Island, in 1844, the son of John Baker, one of the Loyalist pioneers of the district. He took up carpentry as a young man and travelled widely throughout the western United States and British Columbia before settling down in his native province. It is in Summerside and the western half of the Island that his greatest architectural achievements can be found. He designed a number of fine churches, among them St. Mary's Anglican (1907), as well as some of the best business blocks in Summerside, including Holman's and Strong's. The Prince County Hospital and the Capital Theatre were also built from his plans. Mr. Baker also

Longworth House, Charlottetown, Charles Chappell, architect, 1898. - Scott Smith

designed some private residences, including the gracious Sacred Heart parochial house (1894) (see p. 158), just outside of Alberton. He died in December, 1928.

Charles Chappell

Charles B. Chappell was born in Charlottetown in 1857, and during a career that spanned 45 years he gained a reputation as one of the most prolific and prominent Island architects. He formed associations with John Lemuel Phillips and later, John Hunter. Chappell and Phillips worked extensively throughout the Maritimes. Chappell's earliest works were the Charlottetown City Hall (1887) and many of the buildings on Richmond Street after the Great Fire of 1884. His later works included Zion Church, Longworth House (in association with Elliot and Hopson, Halifax, N.S.) (see above), Prince of Wales College (1898), the Queen's County jail (1907) and many private residences in Charlottetown, including his own at 80 Euston Street. Mr. Chappell died in 1931.

John Corbett

John Corbett was born in Ireland c.1827. He immigrated to Memramcook, New Brunswick, and in 1866 moved to Charlottetown, where he remained for nearly 15 years. Mr. Corbett was a mason-architect whose training was practical rather than academic. He was the preferred mason-architect of buildings associated with the Roman Catholic Church and superintended the construction of such buildings as the Bishop's Palace on Great George Street (1872-1875) (see p. 57), the Convent of Notre Dame (1870), St. Patrick's School, the Welsh-Owen Building at 47 Queen Street (c.1872) and the Summerside jail. He favoured substantial brick structures with a minimum of decoration and is credited with the design of St. Patrick's Roman Catholic Church in Fort Augustus (completed 1870, destroyed by fire 1897) and St. Martin's Roman Catholic Church in Cumberland (1868). Mr. Corbett died in Ottawa in 1887.

William Critchlow Harris

William Harris was born in Liverpool, England, in 1854, and at the age of two immigrated with his family to Charlottetown. He apprenticed in Halifax with architect David Stirling, who taught him the fundamentals of the early Gothic Revival. During his career, Harris designed or contributed to the design of some 120 buildings throughout the Maritimes between 1880 and 1913. Perhaps best known

Caroma Lodge, Charlottetown
- photograph by Robert C. Tuck

for his church designs, Harris' interpretation of the High Victorian Gothic Revival vocabulary is unmistakable throughout the Island, and indeed the Maritime landscape. He also designed many fine houses in the region, perhaps the earliest and most famous of which is Beaconsfield (1877) in Charlottetown (see p. 114). Other noteworthy Island houses that he designed are: the Richard Young house (1887) on West Street (see p. 35), the Dundas Terrace (1889) on Water Street, "Hazeldean" (1895) in Springfield (see p. 108), Caroma Lodge (1895) on Grafton Street (see above), H.C. Mills house (1898) on Central Street in Summerside (see p. 83) and St. Paul's rectory in Charlottetown (1889).

William Harris was an architect of considerable dimension. He possessed a particular perception of and sensitivity to the architectural and cultural conditions of his own time and place. He died in 1913.

John Hunter

John M. Hunter was born in Lanarkshire, Scotland, in 1881 and received his architectural training in Glasgow. He apprenticed with an architectural firm in Montreal that specialized in church design. One of his first and foremost architectural accomplishments was the design for the reconstruction of St. Dunstan's Basilica in Charlottetown (1914-1919). Shortly after its

St. Brigid's Roman Catholic Church, Foxley River, John McLellan, architect, 1873.

completion, he formed a partnership with Charles Chappell and this firm carried out many significant projects in the ecclesiastical, hospital and residential fields. They designed a new Charlottetown Hospital, the Summerside High School and the reconstruction of St. Mary's Roman Catholic Church in Souris (1928-1930). Chappell and Hunter was one of the first Island architectural firms to specialize in renovations and designed additions to St. Dunstan's University and the nurses' residence at the Provincial Sanatorium in Charlottetown. John Hunter died in 1942.

John McLellan

John McLellan was born on Prince Edward Island in 1820. Primarily a designer of churches, Mr. McLellan studied under P.C. Keely in New York. He was involved in the interior design of Keely's church at Tignish (St. Simon and St. Jude) and contributed to the design of the nearby convent and parochial house (see p. 158). McLellan executed a considerable amount of work in the Diocese of Charlottetown during the episcopacy of Bishop Peter McIntyre. He designed St. Joseph's Convent and Roman Catholic churches at Foxley River (St. Brigid's, 1873) (shown above) and Vernon River (St. Joachim's, 1877). There is evidence to suggest that he was involved in the design of the Church of Christ in Montague (1876) as well. Mr. McLellan died in Sydney, Nova Scotia, in 1887.

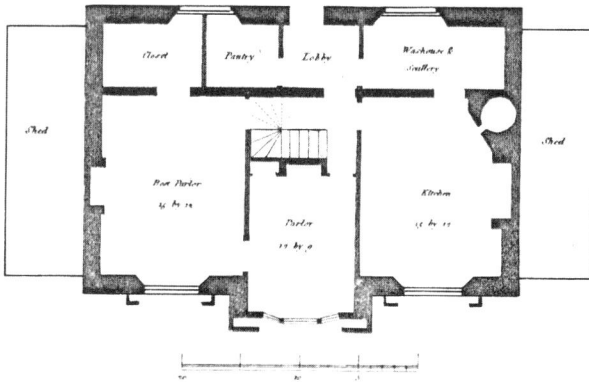

from Sketches for Country Houses *by John Plaw*

John Plaw

John Plaw was one of the earliest and most influential architects to arrive on Prince Edward Island. While none of his buildings remain, his writings and tutorials left a lasting impression on the Island's young builders and architects.

Plaw was born in Putney (London), England, in 1745. He apprenticed for nine years and received a number of recognitions and honours for his work. He developed a specialty in circular country villas and in the late 1700s published three significant books on rural architecture: *Rural Architecture* (1785), *Ferme Ornée* (1795) and *Sketches For Country Houses, Villas and Rural Dwellings* (1800). These books were widely read, not only in Britain and continental Europe but in North America as well.

His apparent discontent with the architectural profession in England and his understanding that Prince Edward Island was a colony in urgent need of public buildings motivated his decision to immigrate to the Island in 1807. He immediately submitted plans for a new jail (never built) and a new courthouse, built on Queen Square in Charlottetown (1811-1814) and demolished in 1972. Perhaps his best-known building was the Round Market, built in 1823, three years after his death, on Queen Square by Island builders Isaac Smith, Henry Smith and Thomas Hodgson. It was moved off the Square in 1842 and replaced in 1867.

Isaac Smith

Isaac Smith was born in the small North Yorkshire town of Helmsley, England, in 1795. When he was 22 years old he immigrated to Prince Edward Island with his new wife and his younger brother Henry in search of a more promising future. Isaac and Henry were trained carpenters, and their first major commission was to build the Round Market on Queen Square in Charlottetown in 1823, to a design

Above left, Fanningbank, Charlottetown, 1834, Isaac Smith, architect. Above right, Isaac Smith in later life. - PAPEI

by the aforementioned John Plaw. By the 1830s Smith was extremely busy designing and building several courthouses, jails, churches and private residences throughout the Island. In 1832 he received his next major commission — to design and build, with the help of brother Henry and Nathan Wright, an official residence for the Lieutenant-Governor. By 1834 the elegant "Fanningbank" (see above) was completed, and today it still commands a scenic view overlooking Charlottetown harbour. As Smith's reputation spread, many more projects followed: bridges, schools, a stone house called Barton Lodge (c.1846) on the Winsloe Road (see p. 36), Prince of Wales College (1860) and Point Prim Lighthouse (1845). The most famous Isaac Smith building was built between 1843 and 1848. The Colonial Building, later known as Province House, was the epitome of the British Classical tradition. As the site of the historic

Confederation Conferences in 1864, it has a permanent and significant place in the history of our nation.

Smith was a meticulous and frugal builder who took a great deal of pride in his work. He had always referred to himself as a builder, but upon the completion of the Colonial Building, in the census of 1848, he called himself an "architect." Perhaps he saw the Colonial Building as the culmination of his career. A devout Methodist, Smith abruptly decided to devote the next fifteen years of his life to travelling the Atlantic region in the service of the Bible Society. He eventually settled in Maitland, Nova Scotia, and died there in 1871.

United Services Officers Club, (Lowden House, Charlottetown)(1869), David Stirling architect.

David Stirling

David Stirling was born in Galashiels, Scotland, in 1822. The son of a stonemason, he received architectural training in Scotland before immigrating to North America around 1847. In search of a livelihood, Stirling lived alternately in St. John's, Pictou, Toronto and Halifax, where he eventually established an office and a reputation. He and his partners were responsible for many government projects in Nova Scotia but also executed plans for the Bank of Prince Edward Island in Charlottetown (1867), the Hyndman Building (1866), the United Services Officers Club on Haviland Street (1869) (see above, p. 117) and the Queen Square Court House and Post Office, built in 1872 and later destroyed by fire.

A declining economy and fewer commissions, along with his success in the competition to design the new "lunatic asylum" in Charlottetown prompted his relocation to Charlottetown in 1877. It was his partnership with the young William Critchlow Harris, however, that led to an increasing number of commissions. Harris had apprenticed in Stirling's Halifax office between 1870 and 1875, so the two architects formed a cohesive Island team. Their most notable projects were the Kirk of St. James, Charlottetown (1877), and Tryon Methodist Church, Tryon (1881). They also supervised the construction of federally-designed public buildings in Summerside, Charlottetown and Montague. In declining health by the 1880s, Stirling relinquished many of his firm's responsibilities to his partner. He died in Charlottetown in 1887.

In Tignish, on the left the house of Sir Charles Dalton, co-founder of the Island's fox farming industry; on the right, a building that was initially Dalton's Drugstore.
- sketch by Robert C. Tuck

Percy Tanton

J.W. Percy Tanton was born in Sherbrooke, Prince Edward Island in 1868. He learned the building trade from Allan Sharpe in Summerside and studied architecture in New Brunswick at age nineteen. He became a prolific contractor in the Summerside area and western Prince Edward Island. (see p. 67) Among his most noteworthy projects are cobblestone houses in South Kildare (Raynor, 1916) and in Summerside (Bowness, 1906), the Dalvay Hotel (c.1901) in Tracadie, the F.W. Strong and Brace and MacKay brick stores in Summerside and the Dalton and MacArthur blocks in Summerside. He is responsible for the construction of the railway stations on the Murray Harbour branch line and has been credited with designing some of the Island's stone cottages. Tanton designed and built many private homes and churches in Summerside and built the original Dalton drugstore in Tignish. (above)

Mr. Tanton also built a Presbyterian church in River Charlo, New Brunswick early in his career, company houses in Sydney, Nova Scotia in 1899 and several buildings in Vancouver in the early 1900s. He helped to form the M.F. Schurman Co. Ltd. in Summerside, operated a lumber mill and bred silver foxes from 1911 to 1917. He died in Summerside in 1937.

Highlands interior, Georgetown.

MacDonald-Mair House, Georgetown.

Gallant House (Barachois Inn), South Rustico.

- Scott Smith

Fairholm, Charlottetown.

Hazelgrove.

Witter-Coombs House, Charlottetown.

- Scott Smith

Beaconsfield, Charlottetown.

Victoria.

- MRMS

Prospère Gallant House, Abram's Village.

Verandah of abandoned house.

The Smith House in Kinkora, an area of Irish settlement, was built in the 1870s by Ambrose Monaghan.

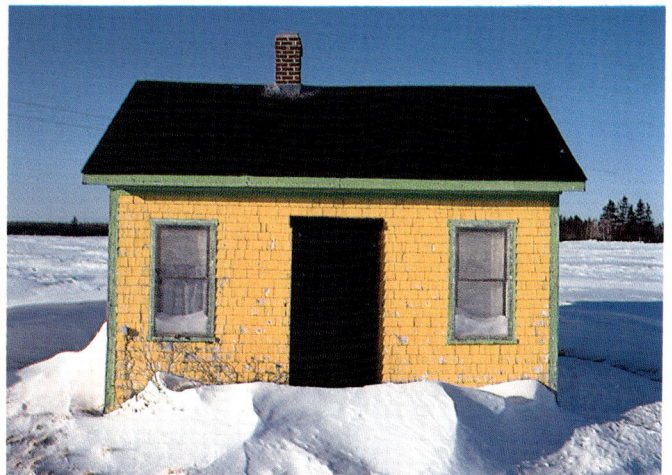

Small house in Coleman.

Ramsay House, Beech Point.

Window detail, Pownal.

- Scott Smith

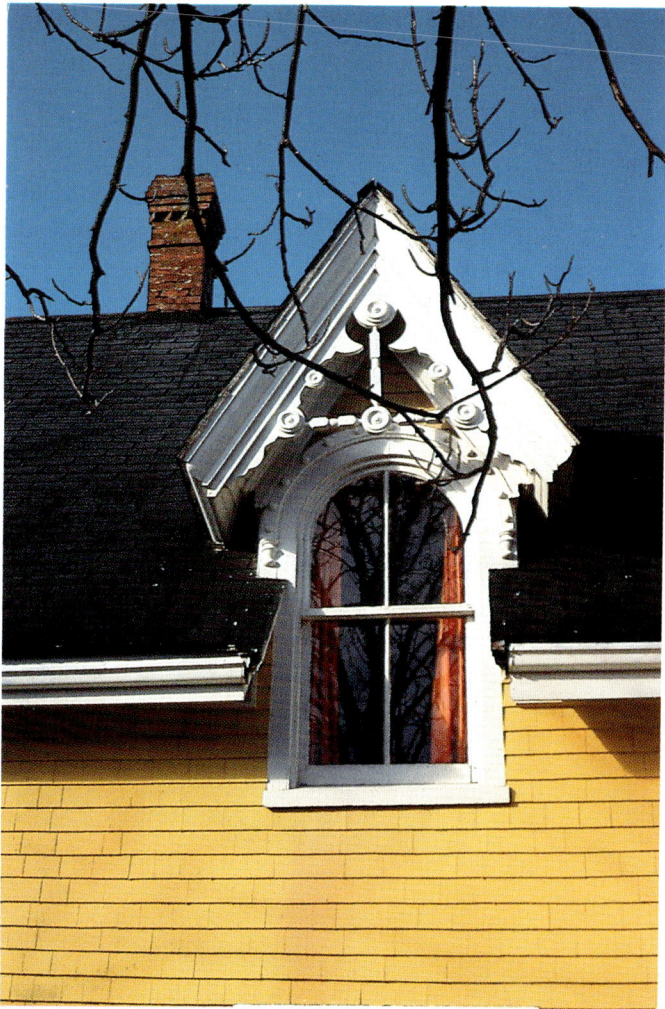
MacLeod House dormer, New Glasgow.

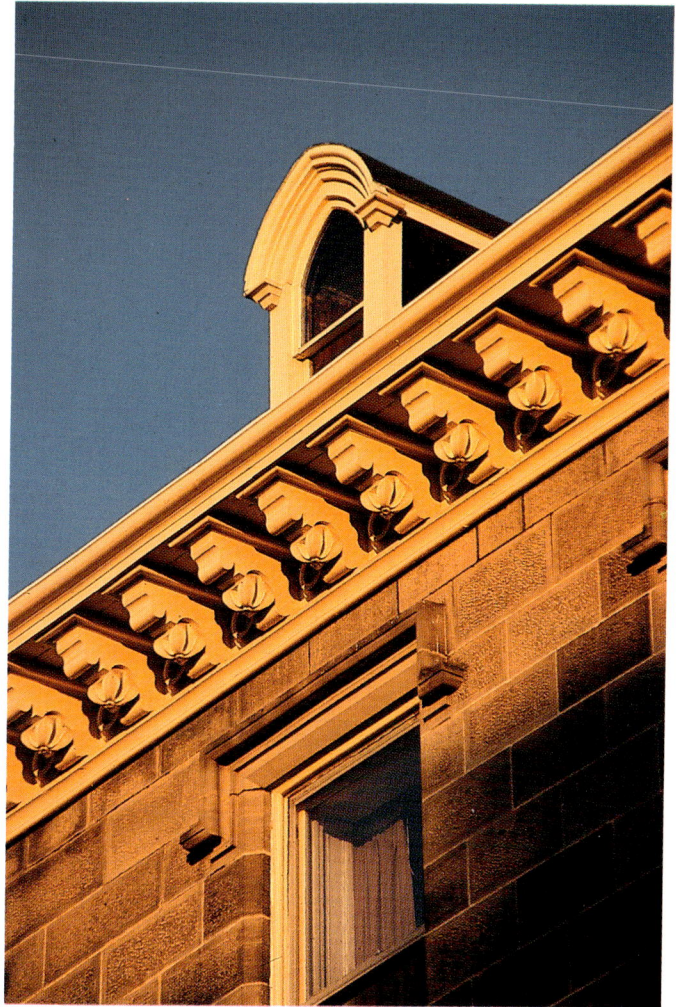
Bishop's Palace, Great George Street.

Fairholm entrance hall, Charlottetown. - Lawrence McLagan

Goff House, herringbone fence, Woodville Mills.

Aitken House sandstone, Lower Montague.

Front door, MacNeil House, Tyne Valley.

McCabe House, Middleton.

Atwell House, Clyde River.

- Scott Smith

The Historic Houses of Prince Edward Island

I am convinced that ignorance, or more accurately innocence, is important in design. One should not know too much. Experience is something else — one can never have enough, one cannot learn the craft of architecture except through experience.

Arthur Erickson

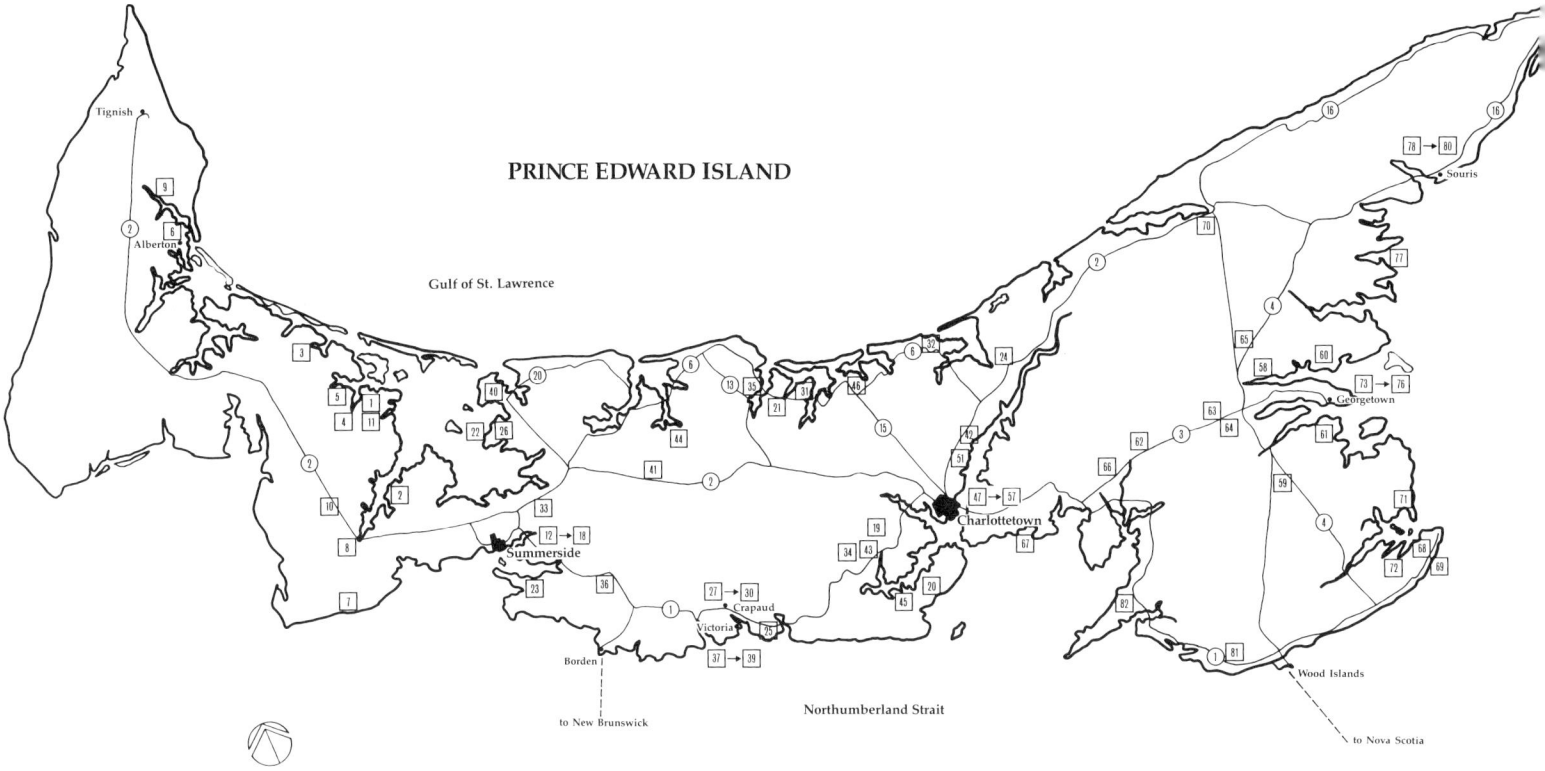

PRINCE EDWARD ISLAND

Gulf of St. Lawrence

Northumberland Strait

Tignish

Alberton

Summerside

Crapaud
Victoria

Borden

to New Brunswick

Charlottetown

Georgetown

Souris

Wood Islands

to Nova Scotia

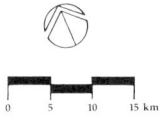

0   5   10   15 km

Location and Index of Houses

Green Park, Port Hill.

"Green Park"
Port Hill

The shipbuilding industry on Prince Edward Island began with a demand for timber in western Europe. Prince Edward Island and other British North American colonies were looked to not only for timber but for vessels designed for the carrying trade. Initially a group of entrepreneurs and shipbuilders from the southern English counties of Cornwall and Devon exploited this market. In the early 1800s they established shipbuilding and lumbering operations on Prince Edward Island. The first vessels, mostly schooners, were laden with timber and delivered to British customers. It wasn't until the late 1830s that some of the builders began to operate vessels themselves, and the industry began to flourish between 1840 and 1889. It reached its zenith in 1865 with a fleet of almost 400 vessels, most of them brigantines and barques and some larger ships. The industry declined rapidly in the late 1800s because of rising costs, timber shortages and the advent of steam power and iron ships.

One of the most prolific and prosperous of these shipping magnates was James Yeo, a native of Kilkhampton, Cornwall, who immigrated to Prince Edward Island in 1819. "The Ledger Giant" was one of his more complimentary nicknames (he was an intense, competitive and often unscrupulous individual). Nonetheless, he was instrumental in establishing the shipbuilding industry in the Port Hill - Bideford area of Prince County. In 1865 his shipyards reached a peak in production: eighteen ships built and sold to British companies. He also managed a fleet of commercial vessels and owned shares in 135 others. His flourishing mercantile trade

Green Park, c.1880.   - from *Meacham's Atlas*

included both coastal and overseas destinations, such ports as Liverpool, Swansea and Appledore, Devon, where his son William was himself a prominent shipbuilder and importer.

After the death in 1868 of James Yeo Sr., his sons James and John continued the shipbuilding operation. James Yeo Jr., born at Port Hill in 1827, set up his own shipyard at the head of Campbell Creek. He also built a fine house and stores there and called the place "Green Park." Between 1856 and 1886 he launched at least 23 of the largest and finest ships that the Yeo family was ever to build. He was also active in politics and from 1876 to 1891 served as a Federal M.P. for Prince County.

His magnificent house, built by James Campbell, certainly befitted his substantial economic and social stature. It is Gothic Revival-style architecture at its finest — a tall, central gable with decorative bargeboard that seems to dominate the front elevation. A beautifully detailed octagonal cupola was built at the intersection of front gable and main roofline to monitor shipyard activities. The main frontal section of the house is well proportioned due to a verandah that wraps around the front and sides of the house. The windows are also well planned, with those on the first floor having nine-over-six sashes, while the second-floor windows are round-headed with pediment entablatures and the third-floor windows are round-headed but with four-over-four sashes. The rear kitchen wing, however, is proportionally different and detracts somewhat from the sophisticated lines of the main house. An enclosed front vestibule has been removed. Inside, the kitchen hearth has been fully res-

FLOOR PLAN: *Green Park, Port Hill*

Octagonal cupola, Green Park, Port Hill.

tored and the black-painted slate fireplace in the parlour, although non-functional, is a tribute to ornate Victorian design. The basement, with its substantial sandstone foundation walls, contains some interesting features — a brick furnace that is still intact and remnants of an old wine cellar.

Built on a beautiful site overlooking the Bideford River, Green Park is now one of the largest provincial parks on the Island. The barns, granary, carriage house and forge have been replaced with a modern interpretive centre, but the house itself has been successfully restored by the Prince Edward Island Heritage Foundation and the Department of Tourism. After James Yeo Jr.'s death in 1903, his son Collingwood Yeo inherited the property. The Prince Edward Island government acquired the house and property in the early 1960s, after it had been unoccupied for nearly twenty years. Restoration began as early as 1968, when the museum was opened within the house. By 1973, Centennial Year, the Interpretive Centre

had been opened and the house had become a fully restored and refurnished memorial to the lifestyle of its builders. Restoration has continued with the refurnishing of the third floor in 1974-75 and the reconstruction of two barns in 1976. The well, with its canopy and windlass, remains intact and functional, but the giant imported English willows and rose garden, symbols of a more affluent period, have disappeared from the grounds. Green Park has become a branch museum of the Prince Edward Island Museum and Heritage Foundation and a popular tourist attraction in the summer months.

MacLean House, South West Lot 16, Prince County.

MacLean House
South West Lot 16

Built in a somewhat exaggerated Queen Anne style, this house near Grand River, Prince County, was actually designed by Percy Tanton, a carpenter and self-taught architect from Summerside. (see p. 48) It was built around 1912 by Willie James Ellis, a master carpenter and shipbuilder from the Northam area, for the newly wed James MacGillivary MacLean, a fox rancher and oyster farmer. The house was subsequently owned by his son Garth, who had to extensively repair the open polygonal turret. It was beginning to settle and the foundation required reinforcement. The flooring and balustrade of the second-floor balcony had to be replaced as well.

Presently the house is owned by Garth's son Dwight and his wife, Jean. They have repainted the exterior and are in the process of restoring the interior.

The front facade is an apparent attempt by the architect to establish a proportionate balance of elements. The Queen Anne style, however, is essentially asymmetric and any attempt at modification needs careful consideration and calculation. The MacLean house, nevertheless, is a rare example of an architect-designed house in rural Prince Edward Island from that particular period.

Williams House, Poplar Grove.

Williams House
Poplar Grove

It is quite evident that skilled hands were involved in the construction of George Williams' fine ell farmhouse in Poplar Grove. In fact George is descended from a long line of shipbuilders and joiners. When William Ellis, a master shipwright from Bideford, North Devon, England, founded the Ellis and Chanter shipyard at Bideford, Prince Edward Island, in 1814, he brought with him a skilled cabinetmaker and master carpenter, Edward Williams of Appledore, Devon, George's great-grandfather. It was Edward's son Robert, himself a farmer and joiner, who built this fine house and barn around 1865, but it was Robert's son and George's father, Harry, who was the post prolific builder of the three. Harry

Sawtooth frieze and window hoods, Williams House, Poplar Grove.

Wellington Williams, or "Little Harry" as he was called, apprenticed under George Gard in Mill River for two years and was paid a princely sum of $2.75 a month! He then went to school, studying blueprint reading and drafting at night and architectural design at a school in Mas-

sachusetts. Upon his return he designed and built many fine houses and churches in Prince County. He was a painstaking and thorough builder who ensured that the job was done correctly and the building would last. When he chose to embellish, such as in the splendid St. John's Anglican Church in Ellerslie, he did so with taste and integrity. He also learned a lot from his uncle Robert Douglas Ellis and from his father, a master in frame cutting. Harry would cut and assemble his frames on the ground, often with dovetailed bracing, and when hoisted into place they would fit with great precision. He also developed a clever horse-powered saw gear that attracted a great deal of attention. Little Harry died in 1934 but left behind a legacy of fine buildings that his son George thinks are a fitting memorial to his work. George has kept his father's saws, planes, mouldings and brass-cornered level, and is very proud of the steadfastness, determination and perfectionism of his ancestors.

The detailing and craftsmanship of the Williams house are quite remarkable. The sawtooth frieze decoration, dentils, window hoods and corner eave brackets give the facades contrast and depth. The detailing in the gingerbread, frieze decoration and column supports of the front verandah are truly a masterpiece of design and finish carpentry. The traditional ell farmhouse has, in this case, been elevated in stature through the design sensibilities and skill of Robert Williams and his son, Little Harry. George lives in a St. Eleanor's nursing home now, but the house has been passed on to the next generation. His son Robert and family maintain it with great pride.

Forbes House
Tyne Valley

Forbes House, Tyne Valley.   - Scott Smith

This house was built in 1883-84 by Tom Gamble, a carpenter from Union Corner, for a merchant named James MacMurdo Forbes, whose family were Loyalists that settled in Lot 16 in the late 1700s.

James learned his trade at the Pound Carriage Shop in Margate and opened his Tyne Valley store in partnership with James Coles of Bedeque. The old brown store still stands at the driveway entrance. James' brother, Donald N. Forbes, soon replaced Mr. Coles as an operating partner and took up residence above the store with his bride, Mary Jane Yeo, the daughter of shipbuilder and politician

James Yeo, Jr. In 1887 the partnership was dissolved when James abandoned the mercantile trade in favour of the ministry. James died in Cleveland, Ohio, the location of his last United Church parish. Donald and his wife subsequently moved into the house and had a large family. Donald operated the store until his death in 1918. His widow continued to reside there until her death in 1954.

The detailing on the house is a tribute to both Tom Gamble and Willie James Ellis, a local carpenter who built the kitchen wing for Donald Forbes. The eave bracketing and frieze dentils provide a tasteful transition between

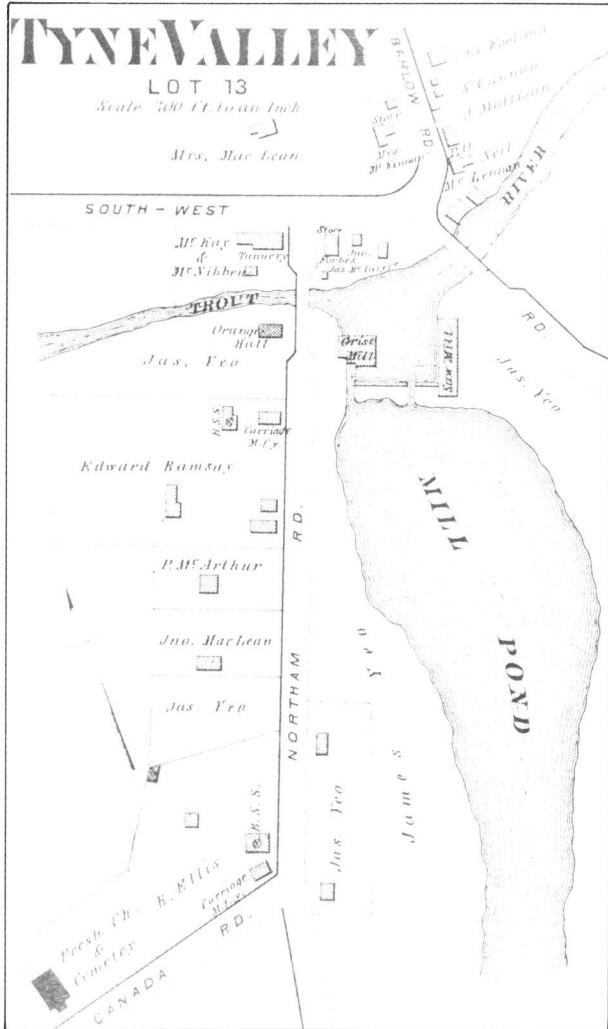

- from *Meacham's Atlas*

Forbes House, Tyne Valley.

the clapboard siding and the bellcast mansard roof. A round-headed second-storey window that breaks the eave line is a relatively sophisticated design element not often seen in rural Prince Edward Island.

The present owners, Nan and Bob Kernaghan, bought the house in 1983 and converted it to an inn. They have added a sunporch/dining room and a carport, but the essence of the Forbes House remains intact – a major contributor to Tyne Valley's considerable charm. (see also back cover)

Pope House, Bideford.

Pope House
Bideford

This elegant ell house has been a manse through most of its history, although it was built c.1870 as a private residence for Thomas H. Pope, an accountant and telegraph operator. He sold it shortly thereafter to the Methodist Church for use as their parsonage. It was during this period that the renowned Island author L.M. Montgomery boarded there for a year while teaching at the nearby Bideford school in the mid-1890s. At Church Union in 1925 it became a United Church manse and remained so until 1972, when it was purchased by Leon Ensalata, who is in the process of restoring and redecorating the house.

Inside, the main staircase and balustrade had to be rebuilt and countless layers of wallpaper stripped. The kitchen and bathroom areas have been slightly altered, but the basic planning remains the same. On the exterior, a troublesome bay window was removed long ago from the front elevation, but the house still retains an elegance that is attributable to the restraint of the decorative elements. The window hood mouldings and the detailing in the vergeboards and column supports of the front verandah are quite refined and give the house a dignity befitting its past.

Callaghan House, Alberton.

Callaghan House
Alberton

Joseph Callaghan's old house on Albion Street, just across from the train station in Alberton, may have been one of Prince Edward Island's earliest duplexes. Built about 1872 by an Acadian, Avie Gaudet, it was purchased in 1918 by Mr. Callaghan from Hardy Oliver, the son of John Oliver, who owned the house before him. Joseph Callaghan shared the house for a time with Otto Smith, and it was probably during this period that the house acquired its duality — the twin gabled roof dormers and two separate front entries.

Mr. Callaghan was a long-time mailman in Alberton who also used to cut block ice from a pond behind the house. Advancing age forced him to abandon the house in 1973. It is presently owned by his son Lawrence, who resides in Virginia but visits regularly in the summer. The house is in urgent need of repair, particularly to the foundation.

Allen House, Union Corner.

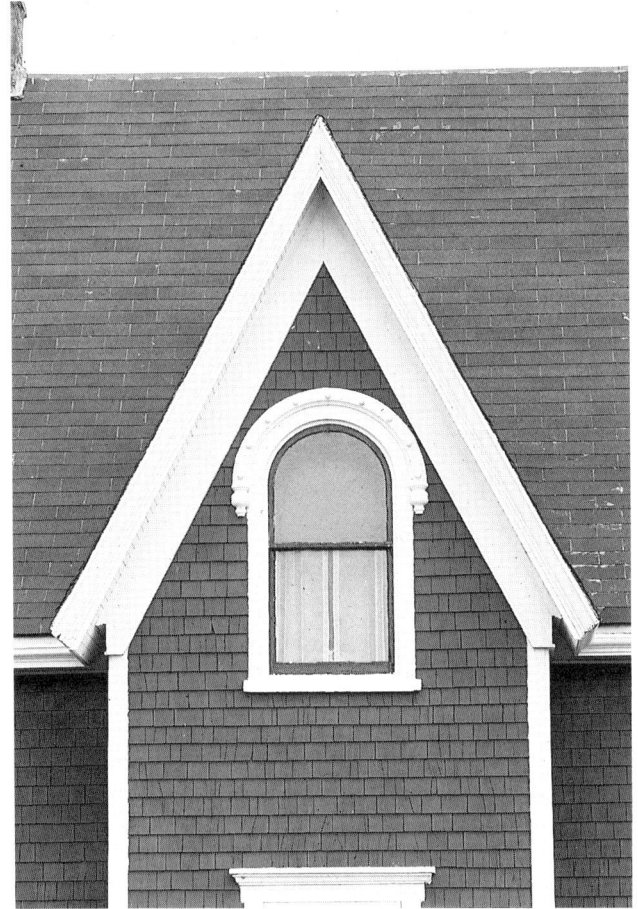

Dormer, Allen House, Union Corner.

Allen House
Union Corner

The Allen house was built about 1910 by Charles Allen's father. The original farmhouse, built by a family named Gallant, is an outbuilding directly behind the house. The kitchen wing of the old house has simply been transferred to the back of the Allen house, a colourful and uncomplicated building typical of many others in the area. The front gable, with its round-headed window and label surround, gives a prominent central focus to the house. Charles lives in nearby Wellington now and his daughter and her husband occupy the house.

Arsenault House, Wellington.

Arsenault House
Wellington

The gaiety and colour of Acadian architecture is quite evident in this fine house in Wellington. It was built in 1897 by Joseph Felix Arsenault to plans drawn by Dr. Andre Gallant. "Joe Felix" was the son of Senator Joseph Octave Arsenault and the brother of Aubin E. Arsenault, Premier of Prince Edward Island from 1917-1919. When his father was called to the Senate in 1895, Joe Felix took over the operation of his general store in Wellington. In 1907 Joe Felix and his family moved to the United States and the house passed through a number of hands before Napoleon Arsenault acquired the house in the mid-1940s. He died shortly thereafter, but his son, the late Sylvere Arsenault, and his family lived there from 1947 to 1986, when his widow sold the house to Theodore and Claudette Theriault.

The house is a joyous festival of bright Acadian colours. It is basically yellow with green trim, but dramatic bands of vibrant red shingles encircle the building above and below the first-floor windows. It is a large, rambling house that the Theriaults have renovated extensively. It has not lost, however, the delightful pedestrian scale generated by the multiple textures, colour variations and the multi-paned, ornate windows. (see front cover)

Gordon House, Huntley.

Gordon House
Huntley

This large, multi-gabled "fox house" was built in 1914 by the renowned Wilfred Maynard for George Campbell Gordon, a fox farmer and nephew of the Presbyterian missionary and martyr Reverend George N. Gordon. The late Mrs. Alice Green, author of the Alberton history *Footprints on the Sands of Time* (1980), was the daughter of George C. Gordon.

The exterior of the house had remained virtually unchanged through a succession of seven different owners, but a family named Bowness made several alterations to the interior, including an enlargement of the kitchen. A front vestibule was also removed, enlarging the downstairs hall. The ornate, split staircase to the second-floor hall has been retained as the central focal point of the plan.

This rambling, spacious house, built in a Four Square style with Colonial Revival detailing, has never been particularly energy-efficient. Two years ago a solar heating unit was installed to assist in hot water heating, and the present owners, Rainer and Denise Zenner, have totally insulated the house. They have also enclosed the two front porches and installed a new metal roof.

Located on Route 152 between Alberton and Montrose, the house once had its own ice cream parlour at the edge of a nearby pond, but this has been converted into a small cottage.

Cameron House
Richmond

This peculiar multi-gabled house stands at the side of Route 2 in Richmond, Prince County. It is a very large house of fourteen rooms and two huge kitchens, an indication that at one time it was shared by two families. In fact, the front section of the house stands on a rubble-stone foundation and is believed to be about 150 years old. There are no nails in this front section: the framing members are joined together by wooden pegs. The rear section is more recent, about 100 years old. The front gable, facing southeast and offset at 45° to the other main gables, is an aspect of house construction seldom seen on Prince Edward Island. The third-storey gables contain some interesting round-headed windows.

The earliest known owner of this 76-acre farm was Coleman Cameron. In 1946 the farm was sold to James A. MacNeill, and upon his recent death Mr. MacNeill left the house to his granddaughter, a resident of New Jersey, in the United States.

Cameron House, Richmond.

Since 1980 the house has passed through several hands but was bought in 1987 by Ron and Sylvia Walsh. They have repainted the exterior clapboards and are in the process of extensively renovating the interior.

77

"Woodbine Cottage"
Port Hill

Woodbine Cottage, Port Hill.   - Scott Smith

Leigh Maynard's house in Port Hill is one of the few houses in the area that contains elements of Greek Revival architecture. The main entry to the house is located on a gable end that is adorned by eave returns, corner boards and a well-proportioned front verandah. The window trim on the front facade is most interesting, with shelf heads on the side windows and ogee and triangular hoods above the second- and third-floor windows respectively.

This charming little house was built c.1865 by merchant and court clerk Hugh A. Ramsay. It was initially called "Woodbine Cottage" and has remained remarkably unchanged through a succession of different owners. Leigh Maynard, first cousin of the prolific Prince County builder Wilfred Maynard, acquired the house around 1960. Mr. Maynard now lives in nearby Tyne Valley, but the house remains in the family. It is now owned by his grandson, Brian Newcombe.

Morrison House, Summerside.   - Lawrence McLagan

Morrison House
Summerside

This elegant Queen Anne-style house was
built in 1908, in the aftermath of the Great Fire
of Summerside of 1906. Until 1930 it was the
home of D.R. Morrison, one of the Island's
most progressive and prolific builders and con-
tractors. His firm was responsible for the construction of some of the most ambitious
railroad and utilities projects, housing and public buildings in Prince Edward Island between
1900-1925.

In 1934 this sprawling house was converted
into six apartments, becoming the town's first
apartment building. It remains so today and
has been owned by Owen Kelly since 1970.
Its circular tower, with a conical roof rising
from a delightful wrap-around verandah, is a
landmark on the corner of Church and Spring
streets.

Holman Homestead
Summerside

Holman Homestead, Summerside.   - Lawrence McLagan

The old Holman Homestead on Fitzroy Street in Summerside was built in 1855, originally as a rectory for a Roman Catholic church (the church, St. Mary's in Indian River, was dismantled and hauled to the grounds of the house in 1854). The house was acquired by the well-known merchant R.T. Holman in 1864. The neo-Classic lines of the original parochial house have not been signficantly altered, although the Holmans did embelish the house consideralby and built two additions — a mansard-roofed library and solarium to the east and a large kitchen wing to the south.

The window detailing on the east wing is quite ornate, particularly the arched and round-headed windows overloking the second-storey balcony. On the west wall of the original house is a delicate second-storey oriel window with small leaded panes. The Holman Homestead was noted for its formal garden behind the house.

In 1972 the house was donated by the two daughters of R.T. Holman, Carrie and Gladys, to the Prince Edward Island Heritage Foundation. The east wing was subsequently converted into an apartment and the remainder of the building became a rooming house, renovated in 1975. In 1984 the Holman Homestead became the home of the International Fox Hall of Fame and head office of the Canadian Fox Breeders Association.

Baker-Strong House, Summerside.   - Lawrence McLagan

Baker-Strong House
Summerside

Stephen Baker lived in this elegant two-and-a-half-storey house for only a short time after he built it c.1866. In 1868 Mr. Baker's house saw the formation of the Summerside Baptist Church within its walls. For a short period it served as the American Consulate. It was bought thereafter by a merchant named Neil Sinclair, who was faced with some immediate renovations. Prompted by his wife, he also had a kitchen wing built onto the back of the house to replace an extremely small kitchen within.

The architectural highlights of the house are obviously the front porch and balcony, added to the house around 1915. The twin panelled front doors with delicate latticework on either side and the fine balcony above with its pedimented gable roof form an impressive visual unit.

The Sinclairs' daughter, Mrs. Heath Strong, lived in the house until a year before her death in 1987. Mr. and Mrs. Bill Martin bought the house from her estate and immediately began renovations. Mrs. Martin now operates a successful interior design firm from an office within the house.

Tuplin-Lefurgey House, Summerside.   - Scott Smith

Tuplin-Lefurgey House
Summerside

Built in 1868 by William Tuplin, a carriage-maker from Margate, this fine Gothic Revival house underwent an extensive reconstruction after it was purchased in 1871 by John E. Lefurgey. Mr. Lefurgey was a noted shipbuilder, merchant and Member of Parliament. Among the many changes he made to the house was the addition of the fine cupola at the intersection of the two rooflines, similar to the one atop the Yeo House in Port Hill and surely a shipbuilder's trademark.

One of the more architecturally interesting houses in Summerside, the Tuplin-Lefurgey house on Prince Street has both neo-Classic and Gothic Revival elements. The pedimented entry of the Classical front verandah is surmounted by a semi-circular bay window with curved panes and a delightful small balcony above it. A centre gable with bargeboard trim contains twin round-headed windows, and the octagonal cupola crowns this most impressive vertical assembly of elements.

The rear wing, which at one time housed as many as twelve of William Tuplin's apprentices, was altered along with other sections of the house to accommodate Lefurgey's large family, many of whom continued to live in the 23-room house long after his death in 1891. In 1924 the house was purchased by J.E. Dalton. Along with other modifications, he raised the entire house five feet. During World War II the rear section of the house was converted into four apartments to house servicemen, and in 1966 the building was sold to a granddaughter of John Lefurgey, Wanda Wyatt. The front section, now known as the Lefurgey Cultural Centre, houses art and dance studios and a craft shop. There were many barns on the grounds at one time and an elegant rock garden and statue used to grace the property between this and the neighbouring Wyatt house. (see also p. 27)

Mills House, Summerside.   - Lawrence McLagan

Bowness House, Summerside.   - Lawrence McLagan

Mills House
Summerside

W.C. Harris designed this unique house in 1898 for Holden C. Mills, a seafood exporter and Anglican church warden. Completed in 1899, its twin corner towers with their conical roofs are similar to those of an Anglican rectory in Summerside (also designed by architect Harris in his later style) which burned down in 1906.

This Queen Anne Revival house, on the corner of Argyle and Central Streets in Summerside, was purchased around 1955 by Jack Scott form the estate of Mr. Mills' widow. He promptly converted it into four apartments, but Harris' trademark turrets, elegant second storey shingling and Dutch gables have been well preserved.

Bowness House
Summerside

This beautifully detailed and colourful house on Cedar Street in Summerside was moved from nearby Sherbrooke around 1962 to its present site. An early tenant and possibly the builder was a man named Abe Bowness, who owned the land on which it now sits. It was acquired by Dick Wedge in 1973 and converted into two apartments.

Designed in the Stick style, a variation of the Queen Anne Revival, the front facade is dominated by its twin bays. The detailing is most attractive: inlaid panelling and dentil moulding of the first-floor bay windows and varied shingle patterns, label window moulding and gable gingerbread of the second-floor projection combine to give the building great texture and character.

MacLennan-Hunt House
Summerside

MacLennan-Hunt House, Summerside   - Lawrence McLagan

A transition of architectural styles is quite evident in this fine house on Fitzroy Street in Summerside. Its gabled roof, blocked eave returns, symmetrical facade and pedimented front projection are elements of neo-Classic, Georgian architecture, but the elongated proportions and Gothic door and window over the front entry are indiciative of an emerging Gothic Revival style.

The house was built ca. 1860 for merchant and trader Colin MacLennan. For a short period both Joseph and William Henry Pope lived in the house. The former was a private secretary to his uncle, J.C. Pope, a former premier of P.E.I., and the latter, a Father of Confederation. The house was sold around 1890 to a prominent shipbuilder/merchant/ politician named Richard Hunt. (In 1876, architect William Harris had designed an elaborate Gothic Revival house for Mr. Hunt,

but unfortunately it was never built.) Richard Hunt was very active in local affairs, serving as postmaster, chairman of Summerside and American consul during the late 1800s. His daughters, Mrs. Ethel Mussen and Mary Hunt continued to live in the house after the deaths of their parents. In 1970 Charlotte and Lorne Ramsay bought the house from the Hunt estate. The current owners are Robert Boyer and Ronald Holley.

The house has had few changes over the years, although architect George Baker drew plans for an elaborate Victorian renovation, of which only the sun porch was built. There is considerable evidence to suggest that the rear wing of the house may be an earlier vernacular house. The building was saved from destruction twice - in 1906, during the great Summerside Fire and again from the intentions of a developer in the mid-1980s.

Cass House, East Wiltshire. — Scott Smith

Murphy House, Cumberland. — Scott Smith

Cass House
East Wiltshire

An architect (possibly F.W. Chandler of Boston, Mass.) allegedly designed this Italianate house for William and Mary Cass in 1872. The Cass farm was a highly successful mixed farming operation and their house was, and still is, a local landmark. Rita Cass O'Rourke, the last surviving daughter of William and Mary, sold the farm property to potato farmer H.B. Willis in 1964. After spending twenty summers at the house, meticulously tending the gardens and hedges, she finally relinquished it to the same company in 1976.

The present owners, Barrie and Paula Willis, executed several major repairs in 1979 and in 1986 won a Heritage Awareness Award for their commitment to sympathetic restoration. The round-headed upper windows, an Italianate trademark, and the front portico (a more recent addition) are particularly elegant.

Murphy House
Cumberland

Three generations of Murphys have owned this colourful centre-gabled house on the Island's south shore since its construction c.1879 by Thomas E. Murphy. Family pride has maintained its excellent condition even though it has not been permanently occupied since the late 1950s. Prior to 1930 visiting priests from Charlottetown stayed there overnight, and for a brief period in the 1960s it was a tourist guest home.

The present owner, Thomas W. Murphy, lives and farms on an adjacent property. He acquired the house in the mid-1950s from his uncle Seymour, son of Thomas E. Murphy. He promptly removed a summer kitchen from the north end, but that has been the only major change. The round attic window, intricate bargeboard scrolls and an interesting tri-sectional window with a twelve-pane upper sash create a strong focal point in the front gable.

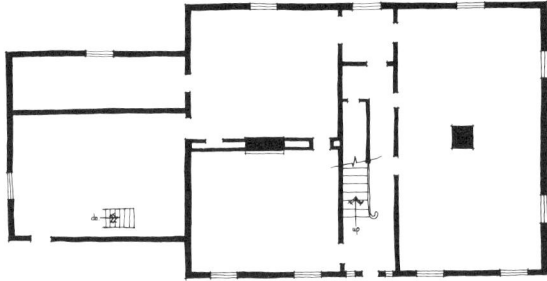

FLOOR PLAN: Gallant House, South Rustico.

Gallant House
South Rustico

Gallant House, South Rustico, before restoration.

This elegant Second Empire house in the historic community of South Rustico is one of the Island's finest examples of a three-storey mansard-roofed building. Its facades are quite refined, with horizontal clapboard siding meeting at tall, slender pilaster corner boards. The smaller eave brackets and six-over-six windows give the front elevation good rhythm, and the central second-floor window has thoughtfully been given sidelights to match those of the front door below. The gabled roof dormers are nicely located in the bellcast roofline, and the front portico is a model of scale and proportion.

The house was built c.1870 by Acadian carpenter Sylvestre Blanchard for Joseph Gallant, a well-known local jack-of-all-trades: merchant, builder, postmaster, farmer and one-time president of the nearby Farmer's Bank. For a while Gallant used half the ground floor as a store, but the building has seen many changes and many tenants have come and gone. Until the 1930s the house was encircled on the south and east sides by a full verandah, and the summer kitchen wing was previously twice its present size.

In 1982 the house was purchased by Gary and Judy MacDonald, who immediately began an extensive and well-planned renovation. The result is a very comfortable and tastefully redecorated six-room guest home called the Barachois Inn. Replete with Victorian antiques and Prince Edward Island art, the Inn offers visitors a reasonably accurate glimpse of Victorian living with modern comforts and accessories. Future plans include a reconstruction of the verandah and the completion of a Victorian garden being presently developed on the grounds.

The MacDonalds' five-year restoration plan has been a costly one, but a successful restoration requires that degree of commitment. The Barachois Inn is a wonderful living monument to an earlier time. (see p. 51)

Ramsay House, Hamilton.

Ramsay House
Hamilton (Beech Point)

This stately sandstone house on the Beech Point Road near Hamilton is slightly larger than (40' x 30') and proportionately different from the other stone houses on Prince Edward Island. In 1974 a wooden kitchen section was removed from the rear of the house. (Traditionally, in many country homes, family life was centred in the kitchen.) As a result of this demolition, the remaining ground-floor spaces had to be completely remodelled and assigned new uses. The basic centre-hall floor plan has nevertheless been maintained. The inside walls were stripped right down to the stone, insulated, waterproofed and refinished. All the window sashes were replaced, and the size of the rear six-over-six windows was decreased with the installation of new stone sills. Some of the rear doors and windows have switched functions because of the removal of the kitchen wing.

The renovation lasted ten months and was very costly. Some serious structural problems had developed on the gable-end walls and iron tie-rods had to be installed at the second-floor level to keep the wall from bowing out further.

Ramsay House, Hamilton.

Over the years there have been other significant changes. The large upstairs hall at one time encircled a winding staircase. This has been replaced by a straight rising staircase, and an additional bedroom has been built at one end of the upper hall. A wooden sun porch that carried a balcony above was removed from the front elevation in 1980. It was preceded by a much larger verandah. Matching brown shingled dormers now grace the front and rear rooflines. Two flues have been removed from interior partitions, eight feet from the north and south exterior walls, but there are no fireplaces; stoves were used until a new fireplace and flue were installed on the north side.

The stone for this house was apparently quarried in the northwest corner of the farm, and the house contains some attractive and unusual masonry, particularly in the front (east) elevation. At ground level, the first course protrudes about three inches beyond the face of the wall, and two feet above that a small striated seven-inch string course runs the length of the front of the house. It too extends three inches beyond the wall surface. The result is not only unique but highly effective in reducing the proportions of the house.

The Ramsay House has always remained in the Ramsay family. It was built in 1855 by Norman and Hugh Ramsay for their father, Archibald Ramsay, himself the fourth son of John Ramsay, who emigrated from Scotland in 1770. Inhabited by a number of Ramsay descendants since then, it now belongs to Norman and Shirley Ramsay. (see p. 56)

Affleck House, Lower Bedeque.

Affleck House
Lower Bedeque

The Affleck house in Lower Bedeque was built in 1826-27 by George Price. The twin two-storey bay windows were not part of the original house, and there have been other alterations to the interior. The old hand-sawn clapboards remain, however, and the house still retains its generous proportions and steeply pitched gabled roof.

George Price sold the house to a man named Peacock, who in turn sold it to the Waugh family. In 1940 the farm was purchased by Douglas Affleck, and in 1979 his son Stewart inherited the house and farm operation.

89

Glenaladale, Tracadie

"Glenaladale"
Tracadie

"Glenaladale" was built on a very historic site near the original homestead of Captain John MacDonald of Glenaladale, Scotland. In 1772 MacDonald chartered a vessel called the *Alexander*, which set sail for Prince Edward Island with 210 emigrants from the Scottish Highlands and the Hebrides. MacDonald himself arrived on the Island the following year and very quickly established himself as one of the most prominent proprietors in the early period. He and his sons owned nearly 40,000 acres in Lots 35 and 36.

Captain John MacDonald's grandson, Sir William MacDonald, was born in 1831 in a small frame house quite near Glenaladale, on the bank of Tracadie Bay. Sir William was later to go on to great things. He was the founder of the MacDonald Tobacco Company and MacDonald College near Montreal. He became a very wealthy man and a leading philanthropist in the field of education.

This substantial brick house was built in 1883-1884 by Sir William MacDonald for his brother John Archibald MacDonald. It was designed with Georgian proportions by a Montreal architect and built by contractor James Hodgson. Henry J. Cundall, a prominent merchant and real estate agent in Charlottetown, acted as building superintendent for Sir William, who was living in Montreal at the time.

Sir William wanted an Island house for his brother that would last a long time and be fire- and frost-proof. No expense was to be spared, and the estimated cost of $9,000 (final cost $9,500) included windows, doors, mouldings and other finish woodwork imported from Montreal. The framing lumber and flooring, however, was milled by William Haywood of Lot 5 in Prince County. The outer and innter walls, 21 and ten inches thick respectively, were made from brick that had been rescued from an older brick storage house which stood near the dock at the water's edge. Although this warehouse burned down, the bricks, made in a kiln on the property, survived in excellent condition. Other Island-made brick was used to complete the house, which sits on a foundation of brick and Island sandstone. The slates for the hipped roof were sent by ship from Montreal to Charlottetown and by train to Tracadie.

The house is essentially (50 feet) square (see floor plan), and the spacious rooms are arranged around a wide central hall and open

90

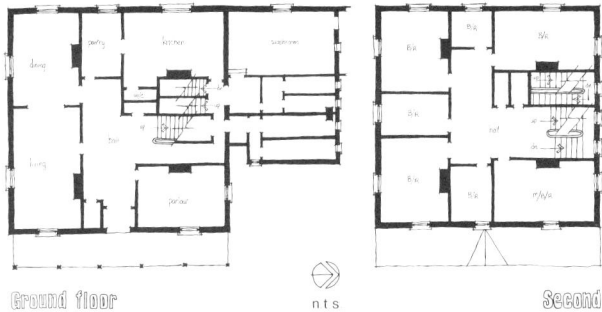

Ground floor   nts   Second

FLOOR PLAN: Glenaladale, Tracadie

Glenaladale c.1910.   - PEIMHF

staircase, an apparent oversight in fire protection planning. Apart from the installation of electric power and a bathroom on the ground floor, there have not been many significant changes to the house. There is a central floor furnace, and the ingenious spring-fed water system has been extended mechanically only to the downstairs kitchen and bath. The front verandah has been altered somewhat over the years but still retains its restrained, dignified lines.

Almost as impressive as the house was the large barn, built concurrently nearby at a cost of $22,000 and connected to the house by a long covered walkway that could be disassembled in case of a fire in the barn. In fact, the brick termination of this walkway saved the house when the barn was ravaged by fire in the fall of 1907. What a splendid barn it was! Despite Henry J. Cundall's recommendation to build a courtyard-type barn complex, Sir William erected a long barn, 300' x 52', with a manure house extending from its north side 75 feet. Horses, cows and machinery were housed on the main floor, with hay and grain

stored on the upper flat and potatoes and vegetables in the basement. The aforementioned water system was also piped into the barn, and fresh water flowed continuously year-round for the animals. The farmyard was also the site of elaborate social events such as the Tracadie Tea of 1887, which attracted a reported 4,000 to 5,000 people. The barn was replaced almost immediately after its destruction by the present barn, c.1910.

Glenaladale was purchased shortly after John Archibald MacDonald's death in 1903 by Charles Major MacKinnon, a prominent farmer, cattle buyer and fox rancher from Rose Hill Farm in Lot 16, Prince County. His stay at Glenaladale was short. In 1911, six years after taking up residence, Major MacKinnon left for the United States. The next year his brother Arthur Wellington MacKinnon took over the farm until his death in 1927. His sons and his daughter Ruth Barlow have lived in Glenaladale ever since.

Redcliffe Downs, Hampton.   - Scott Smith

"Redcliffe Downs"
Hampton

This good example of New England-style architecture was built in two sections by William Inman. The rear section was the original house, built in the 1830s. The front section, added to the original house later that century, overlooks Northumberland Strait. The rear section is now the kitchen wing, and it contains a recently restored stone fireplace (right) that was probably the original kitchen hearth. There is also a fine Palladian window in one of the side gables of the old house. The entire house is filled with period furniture and artifacts dating back to the mid-nineteenth century. A beautifully detailed staircase rises from the central hallway to the second floor in the newer house. The front elevation exhibits neo-Classic elements of nineteenth century New England architecture: the six-over-six win-

Original open-hearth fireplace, Redcliffe Downs, Hampton.

dows have pilaster side mouldings and shelf heads, and the front entry porch is well designed. The four-panel, single-leaf door is framed by flush sidelights and a half-elliptical fan light. Corner pilasters engage the gabled porch roofline in a most harmonious way.

Redcliffe Downs is now owned by Mr. and Mrs. Barrie Willis, who purchased it from David MacKay in 1976. They use it as a summer residence and operate a campground from it.

Simpson House, Hamilton.

Simpson House
Hamilton, (Shipyard Road)

This large, rambling house represents an architectural style that did not emerge as a domestic alternative until the early 1900s. The gambrel roof, a contribution of early Dutch colonists on the eastern seaboard of the United States, had appeared earlier in barn construction.

The house was built in 1900-1901 by the Simpson family to a plan from a pattern book brought east by a brother from British Columbia. He had come to assist in its construction shortly after the original house and barn had burned to the ground. The house has been renovated internally by successive owners, but the basic ell plan and the external features have remained unchanged.

The house and property were purchased in 1978 by Philip Amys Jr. from an American family named Moxley. Mr. Amys and his wife, residents of Halifax, Nova Scotia, now use the house as their summer home. They have added a sun porch across the full width of the rear (south) side of the house.

Inman House (Riverdale), Crapaud.

Inman House ("Riverdale")
Crapaud

On a hill overlooking the Westmoreland River stands "Riverdale," a house of graceful proportions. It was built between 1870, when William Inman acquired the twenty-acre property, and 1882, the year of his death. It has not been possible to affix an exact date of construction.

The house was built in the Picturesque style, a variation of the Gothic Revival movement in the architecture of the day. Its prominent central gable, steeply pitched roofline and a front verandah that runs the full length of the house give it an imposing but well-proportioned front facade. The tall sixteen-pane windows and an entry vestibule beneath the trellised verandah contribute greatly to the scale of the building. A Palladian window in the rear central gable is another fine feature.

The foundation was built with large blocks of hand-hewn Island sandstone, and the basement contains a bricked-in cold storage room with a brick floor. The wood frame is built of hand-hewn cedar beams and posts, and the floors were laid with pine boards.

At one end of the rear entry porch, Mr. Inman built a very interesting addition: a long, narrow (52' x 14') carriage house that also contained a henhouse, privy and possibly a blacksmith's shop. Immediately adjacent to the house there are two sets of sliding double doors, which allowed visitors to drive into the carriage house and alight in shelter in a porte-cochère before the horse and carriage were drawn out the opposite door and stabled. The carriage house also contained an open hearth which may have been used to boil vegetables, scald pigs or as a blacksmith's hearth. This carriage house has been completely renovated for additional living space by the previous owners, Dean and Cathy Shaw.

The floor plan (see p. 95) is of the centre-hall variety and is essentially symmetrical. A gently rising central staircase with curly maple banister and railing ascends to the rear hall, with its Palladian window. The downstairs living spaces have ceilings eleven-and-a-half feet high, and two fireplaces in these rooms are operational.

Riverdale remained in the Inman family until 1975, when Mr. and Mrs. Stephen Gibbons purchased it from Mrs. Colin MacPhail, a niece of Arthur Inman, who had earlier contributed a plumbing system and electrical wiring to the house. The MacPhails also made some restorative effort, refinishing interior walls and instal-

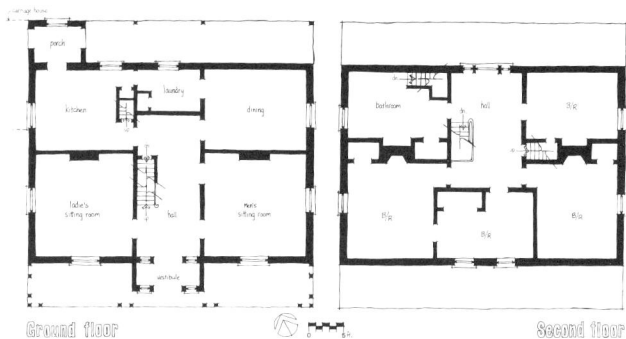

FLOOR PLAN: Inman House, Crapaud

Dawson House, Crapaud.

ling a central heating system. Mr. and Mrs. Shaw purchased the house and property from the Gibbons in 1978, and in 1985 sold it to Jack and Jean Miller.

Mr. Miller, an interior designer, has made substantial changes to the planning and interior decor. A noteworthy removal was a narrow rear staircase that ascended from a small kitchen to what is now a bathroom, but what may once have been servants' quarters. The Millers' sensitivity to colour, finishes and furnishings is impressive, and, above all, the integrity of the original house has not only been retained but enhanced. Future plans include further development of the outdoor living space and surrounding gardens.

Dawson House
Crapaud

This elegant, twin-gabled house was built between 1910 and 1914 by David S. MacQuarrie, who built two other fine homes nearby on what is now the Trans-Canada Highway in Crapaud. Mr. MacQuarrie also operated an undertaking parlour in the house until 1920. He then sold it to Percy Dawson, who with his son Robert continued to operate a funeral parlour into the 1930s. Mr. and Mrs. Rick Holt purchased the house from the Dawson family in 1983.

The Holts have a plan to renovate the entire house and connected barn. Indeed, considerable interior modifications have already taken place. The original link between house and barn is still intact and is quite rare in Prince Edward Island or even Atlantic Canada.

Howatt House
Crapaud

Howatt House, Crapaud.

On the Trans-Canada Highway at the outskirts of Crapaud, in central Queen's County, stands a remarkably well-preserved example of Italianate-style architecture. It was built in 1865 by George Howatt, a merchant whose family were United Empire Loyalists.

The house shows many characteristics of the Italianate movement. Its symmetrical facade, refined proportions and front verandah with coupled pillars and eave bracketing are a delight to behold. The gabled dormer with eave returns built into the front eave line combined with the bracketing give this house one of the most thoughtfully composed and well-proportioned front elevations in Prince Edward Island.

George Howatt operated a general store which stood just north of the house. He converted the third floor of the house into a schoolroom, and a private tutor was hired to teach his children.

In 1883 the property was sold to John Moore, who named it "Orchard Park." Mr. Moore built a track on the farm for local horse racing and was one of the co-founders of the first butter factories in the province, in Crapaud. In 1906 the house passed into the hands of the Simmons family, one of whom, Robert Simmons, was one of the pioneers of the fox farming industry in Prince Edward Island. The property was finally sold in 1970 by Robert's son, John Simmons, to Barry Dawson, who has converted the house into apartments.

Cameron House, Crapaud.

Cameron House
Crapaud

Built in the Stick style, a variation of the Queen Anne Revival, the Cameron House has some interesting features. A pair of two-storey projecting bays dominate the front elevation, and a non-functional umbrage is framed between them, at ground level. The clipped dormer gables form hoods over the second-storey windows and many of these windows have stained-glass panes in the upper sashes. Inside, the wide central stairway is embellished by a beautifully hand-carved banister and newel post.

R.H. Cameron, a prominent merchant, came to Crapaud in 1876 and bought the property from a man named Dawson. The house was built in 1893 by a master carpenter named Nathan MacFarlane for Mr. Cameron. His wages at the time were 60 cents a day and his eight apprentices received 50 cents a day for their labour! MacFarlane built the house on a solid foundation of Wallace, Nova Scotia, freestone. The house was purchased in the early 1900s by the Howatt family, who subsequently relinquished it to the Rogersons. LeRoy Howatt bought the house back in 1945, and his widow, Mrs. Amy Howatt, continues to live there.

Doucette House
Cymbria

Doucette House, Cymbria.   - Canadian Inventory of Historic Buildings

This house may be one of the oldest extant on Prince Edward Island. It is difficult to determine the exact date of construction of a building over 200 years old, but it is estimated that the larger, original section of the house was built around 1773 and that the smaller kitchen wing was added in the early 1800s. It appears that the original kitchen hearth was located in the present living room (see floor plan p. 99).

The Doucette house is a fine example of traditional Acadian domestic architecture. It was built by practical people and there is not an inch of wasted space. The entire building is of a smaller scale than most vernacular buildings; the rooms are tiny, the ceilings barely seven feet high, and the main stair steep and narrow. Beneath the exterior shingle cladding is cedar log construction packed with spruce needles and grass, and the building sits on a stone foundation, one corner of which had originally been sectioned off to form a cold storage room.

There is a minimum of ornamentation in the house. Window and door mouldings are quite plain, and wainscoting is made up of fourteen-inch horizontal pine boards. The ceiling boards are tapered from end to end and fit together with a pleasing angularity.

FLOOR PLAN:
Doucette House,
Cymbria.

Second floor

Ground floor

Orr-Deblois House, Dalvay.

The house remained in Doucette hands until Joseph Doucette sold it in 1982 to John Langdale, the owner of a nearby resort. Until that time, physical changes had been minimal – some interior redecorating and the installation of aluminum windows and doors. The Langdales, intent on making the house their permanent home, have reshingled the exterior, installed a new wood stove and opened up the ground-floor plan. A considerable effort has been made to retain the integrity of the old place, but some changes were unavoidable. The house contains some interesting artifacts and antiques, including an old altar table, a reminder of a time long ago when mass was held in the original dining area.

The Doucette house is a valuable historic resource in Eastern Canada. It tells us of the frugality and skill of Acadian builders and of the practicality and hardiness of the Acadian people.

Orr-Deblois House
Dalvay

In 1898 Colonel A.M. Orr and his wife purchased land quite near Dalvay House at Grand Tracadie, where they had earlier been guests of Alexander MacDonald, one-time president of the United States-based Standard Oil Company. They built "Lakewood Lodge" shortly thereafter and lived there until 1913, when they sold it to some developers from Alberta. George D. Deblois bought the house and land in 1922. His widow, Mrs. Marion Deblois, died in 1983 and left the property to her daughter Helen M. Likely, the present owner.

The house is rambling and eclectic in style, with interesting windows, some containing intricate diagonal tracery. It stands on a large open lot between two small lakes facing the Gulf of St. Lawrence at the east boundary of the Prince Edward Island National Park.

Marchbank House, New Annan.

Marchbank House
New Annan

The original Marchbank house, built in 1825, was a halfway house that offered accommodation and refreshment to weary travellers on their way from Kensington to Summerside. These inns were quite common on the Island during horse-and-buggy days but very few remain. There are many accounts of rollicking good times enjoyed by travellers stranded on stormy nights.

The original house was quite plain, but an extensive renovation in 1926 saw the addition of the beautifully detailed front verandah and balcony. The extended rafters, paired columns and railings of the verandah and balcony give the house an interesting and well-proportioned front elevation. Some of these elements are characteristic of the Stick style, a universal trend in architectural design around the turn of the century.

Percy Marchbank's renovation in 1926 also included the removal of a kitchen wing in the rear; it now sits as an outbuilding behind the house.

When Percy died in 1948, his son John inherited the house. He and his wife have been living there ever since.

Alderlea, Bannockburn Road.

"Alderlea"
Bannockburn Road

The Dixon house, or "Alderlea," as it is known historically, is another basic Maritime Vernacular house type that has been added to and embellished. A lateral wing and a verandah are later additions, and the main house contains some attractive neo-Classic and Gothic detailing, particularly in the front porch and dormer.

The original farm tract of 170 acres near Clyde River was purchased by the Dixons in 1832 from the trustees of the Earl of Selkirk. The house was built shortly thereafter and has remained in Dixon hands ever since. It is presently owned by George and Boyd Dixon.

Toombs House
Rusticoville

Harvey Toombs' house in Rusticoville is a good example of a variation of the ell farmhouse style in which the ell addition is a smaller replica of the main section. In this case, both sections have a tall central gable, but the ell has a front porch which carries its entire length, while the main section has only a small canopy over the front door.

The house was built c.1890 by a carpenter named Harold Gay for Tom Bulman and exists in its original condition except for the changing of a round attic window in the central gable to the present square one. An attractive trellised gate opens into a spacious front yard.

Harvey's grandfather Cyril Toombs successfully operated one of the few remaining family farms in the area. The attractive lines and colour scheme of the house capture the attention of travellers on their way to the Island's north shore.

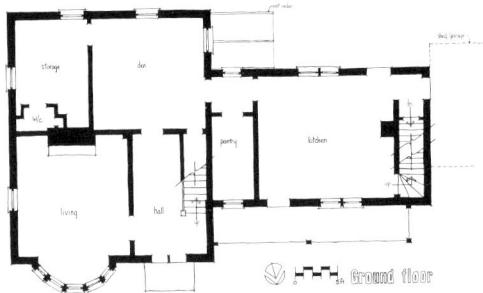

FLOOR PLAN: McCabe House, Middleton

McCabe House, Middleton.

McCabe House
Middleton

Travelling west on Route 1A towards Summerside, one notices a very interesting variation on the traditional ell farmhouse at Middleton. The McCabe house stands at the bottom of a long rise, and attention is at first drawn to the harmonious way that this building has been added to. A rather long storage shed has been joined to the ell section of the house, and the successive lowering of rooflines seems to counterbalance the slope of the surrounding terrain. The colour scheme of cream-coloured clapboards and black roofing is quite dramatic.

Patrick McCabe purchased the original farm property in 1845. He built a small frame house, which has been torn down, but his son Michael built the present house in 1880. The basement was constructed of enormous sandstone blocks, some as wide as five feet. The cost of the work was $1,200, but after the discovery of a construction fault in the foundation, the contract was settled for $1,000! The house remained in the McCabe family, but in 1979 Margaret McCabe found the upkeep of the house too demanding for a woman in her later years and moved to Summerside. Her son George, who lives nearby, tended the house and grounds until 1987, when Lloyd and Linda Larocque purchased it. They immediately converted it to a bed-and-breakfast inn. (see p. 59)

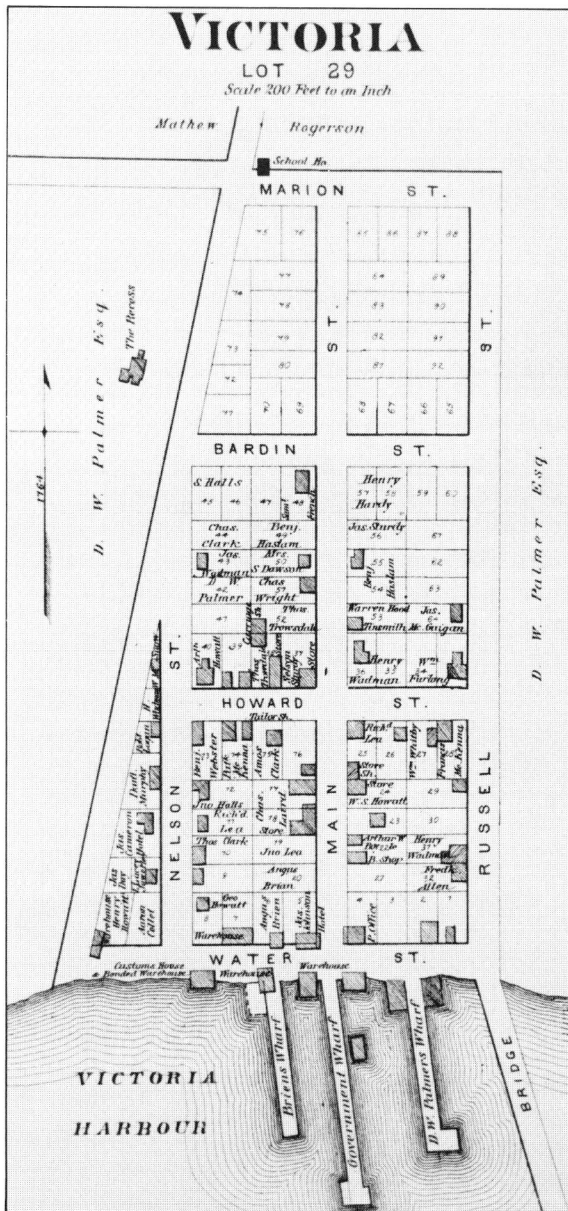

Victoria - from *Meacham's Atlas*

Victoria

Donald Palmer laid out the village of Victoria in 1855 in the proper British rectilinear grid system on a corner of the original Palmer family farm. He recognized its strategic position midway between the important shipping ports of Charlottetown and Summerside and the potential of its sheltered natural harbour. (see p. 54)

103

MacIntosh House, Victoria.

MacIntosh House
Victoria

This beautifully proportioned and colourful one-and-a-half-storey cottage in Victoria was built around 1865 by a family named MacIntosh, who lived in it for more than 80 years. They sold it to a man named Rogerson, whose heir, Jack Lea, sold it finally to the Farrar family of Charlottetown in the early 1960s. Although it has remained unoccupied for the past thirteen years, the exterior, at least, has been well kept. The interior and a rear kitchen wing, however, are in urgent need of repair. It is presently owned by members of the Farrar family, but plans for it are somewhat uncertain.

The most charming features of this house are its scale, pleasing colour scheme and the two gabled dormers with their enclosed round-headed windows.

The Rowans, Victoria.

"The Rowans"
Victoria

The Rowans house has an interesting history. It was hauled across the ice from a site near the People's Cemetery in Hampton in 1881 by John S. MacQuarrie, a shoemaker. It was bought by a Massachusetts man, Edward Archibald, in 1905, and he added the charming wrap-around verandah, front gable and servants' quarters before his son Edward Everett Archibald inherited the place in 1909. Fred Inman bought the property in 1927 and named it "The Rowans" after the rowanberry trees which grace the front yard. Mr. Inman lived in the house until his death in 1970, and for the next ten years his daughters leased the house to tourists during the summer months. In 1980 the house was sold to a Dennison family from Ontario. It is estimated that the original house was built in Hampton around 1850.

Palmer House (Dunrovin), Victoria.

Palmer House ("Dunrovin")
Victoria

One of the more historic houses in this historic south shore town is "Recess III," or "Dunrovin," as it has been renamed by its present owners, Howard and Kathryn Wood. "Recess I" was built in 1827 by James Bardin Palmer, an immigrant Irish attorney and founder of the Loyal Electors, a reactionary political alternative in the early nineteenth century in Prince Edward Island. The Palmers were the main founding family of Victoria, and when the first house was destroyed by fire in 1843, it was rebuilt eleven years later by James' son Donald. Recess III was built in 1910 by Charles Palmer to replace the second house, which had deteriorated beyond repair during his trip to the Klondike. Today it is operated by the Woods as a tourist home.

It is an interesting example of Queen Anne Revival-style architecture, with its two diagonally opposite corner towers and pedimental front portico. The central chimney is cornered on the roof crest to provide more efficient drainage. The house contains some fine old antiques, including the Palmer family's grandfather clock.

Dunrovin sits in a pastoral landscape on the outskirts of Victoria amid outbuildings, tourist cabins and beautiful lime trees. Its rigid weather vanes on the tower peaks can be seen for miles from the various road approaches to Victoria.

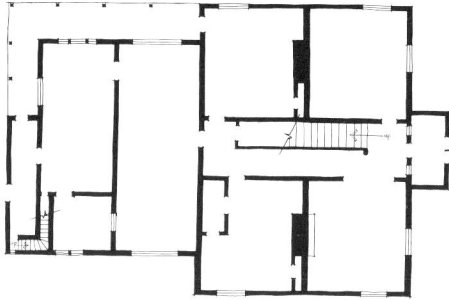

FLOOR PLAN: Keir House, Malpeque.

Keir House, Malpeque.

- Scott Smith

Keir House
Malpeque

The Keir house in Malpeque, now the home of William Auld and his family, has an interesting and significant history. The house consists of two very old sections that were both hauled to the site from nearby locations and joined together. The main front section was built between 1780 and 1800, and the rear kitchen section about 1805. The union of the two took place between 1806 and 1810, when Reverend Dr. John Keir was ordained minister of Princetown Presbyterian (later United) Church. His son William practised medicine from the house, as did William's son James. It was James who built the solid stone walls enclosing the property and a stone well that stands next to the kitchen wing of the house.

A local blacksmith made the wrought-iron gate in the early 1900s. A shingled boathouse that was hauled up from a nearby shore stands at the rear of the property. Its top floor was converted into Mrs. Keir's painting studio.

The main house has some extremely wide pine baseboards, and the interior walls contain some of the earliest examples of plaster finishing in Prince Edward Island. The design of the house contains both neo-Classic elements and Georgian proportions. It is a large, two-storey building with a hipped roof and a symmetrical front facade. The front portico carries a pedimented, enclosed balcony above it.

The house was bought in 1960 by William Auld, a grandson of Dr. James Keir.

Hazeldean, Springfield.

"Hazeldean"
Springfield

This charming house on the Rattenbury Road was designed by Island architect W.C. Harris and built in 1895 by Nathan MacFarlane of Fernwood, who also built St. Mary's Roman Catholic Church in Indian River. Two aspects of the front of the house which immediately catch one's attention are the three gabled dormers, all of different sizes, and the delicately proportioned umbrage with its delightful gingerbread trim. The house also has interesting proportions, with the front eave line lower than that in the rear. Some of the interior woodwork was handgrained by a man named Inglis who also did similar work in St. John's Presbyterian Church in Belfast.

"Hazeldean" is the name given to the farm by George L. Haslam, who bought it from his father, Robert Haslam, and who originally built the present house and outbuildings in a more central location on the farm property. The Haslam family had acquired this property in 1883 from Joshua Murray. After George's death the house was passed on to his son Reginald. His widow, the former Doris Muncey, and his sister Muriel now own the farm.

Tulloch House, Lower Marshsfield.

Tulloch House, Lower Marshfield, c.1880

- from *Meacham's Atlas*

"Tulloch"
Lower Marshfield

This charming house has undergone a rather interesting transformation since its construction c.1820. It was designed for the Hon. Donald Ferguson, M.P.P. and Commissioner of Public Works, as a residence of some grandeur and moderate proportions. However, the mansard roof with its iron filigree around the crest was never built, and a charming front verandah has disappeared. What remains is an interesting study of reduced proportional effect. The windows and doors, except for the addition of wooden storm doors at the front entry, have remained unchanged. Crawling vines have been used effectively to enhance the buildings clean, symmetrical facades. Tulloch still commands an interesting view, but the once impressive grounds have been reduced to a smaller front lawn and the main road access to the property is now from a different direction.

The house remained in the Ferguson family until 1920 when it was purchased by Henry Boswall. It is now owned by his son, Lyle Boswall.

Atwell House ("Tich-Na-Craig" - House of Stone)
Clyde River

Atwell House, Clyde River.

Approaching Charlottetown from the west on the Trans-Canada Highway, a traveller's attention is drawn to a remarkable and unusual stone house at Clyde River. At the end of a long, winding lane stands the house of Dora Atwell, a living testimony to the principles of efficient and sensitive restoration.

The house stands on a property that was originally glebe land granted to Major Ambrose Lane by his father-in-law, Lt. Governor C.D. Smith, in 1837. In 1840 Major Lane sold the property to a Thomas Kickham, who named it "Dog River Farm." He then sold an adjacent farm, and with the money began to build a stone house in 1842, with the help of some local stonemasons by the name of Heartz. In 1843 Ambrose Lane re-acquired the property and used the house as a hunting lodge. A large rock at the rear of the farm was subsequently named "Lane's Rock." The farm became known as "Sherwood Farm" after Lane's death, and a long succession of different owners and tenants followed. After an eighteen-year period, during which the house was used as a grain and vegetable barn, the property was bought by the late Mr. Atwell in 1973 from Harry Boyle and the estate of Robert Boyle.

The arduous eighteen-month task of restoring "Sherwood" was then begun. The ground-floor beams and joists had to be replaced, and the north roof framing had to be rebuilt. The main central chimney (six feet by four feet) and the back porch had to be reconstructed using Island sandstone. The walls, incidentally, are cavity walls, 21 inches thick with an eight-inch air space, and the stone apparently was quarried

Atwell House, Clyde River.

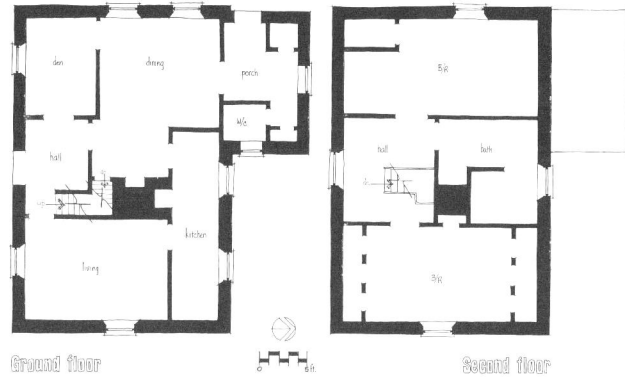

FLOOR PLAN: Atwell House, Clyde River

on farm property. Mrs. Atwell believes that the finish lumber used in the house was returned to Prince Edward Island from England, where it had been sent to be milled. The boat discharged its cargo on a dock, at the foot of the slope below the house, after sailing up the Clyde River.

There has been some conversion of space, particularly upstairs, where additional closet space was required and new bathroom facilities had to be constructed beneath one of the dormers in what was formerly a bedroom. Up until 1973 the house had no plumbing except for a water pump in the kitchen.

The Atwell house is a classic example of the central-fireplace plan popular in the eastern counties of Lowland England and among seventeenth-century New England colonists. The house's orientation is southerly, and, with ten-foot ceilings in the ground-floor living spaces and nine-over-six windows on the southern elevation, the downstairs is bright and roomy.

A narrow (30 inch), winding stairway leads from the entry hall to an upstairs hallway within the "eyebrow" or "nun's hood" dormer.

The historical integrity of the house has been retained in a consistent and commendable way. A fireplace mantel was made from a beam from one of the original barns. The windows in the house still have their original shutters. The wallpaper has been printed from English blocks (Guiness Brothers) that are over a century old and still used locally. Most of the door knobs are the original ones, made of rosewood or mahogany. A Nova Scotia token penny dated 1840 hangs in a glass encasement near the front door lock from which it was taken, used probably as a washer. Mrs. Atwell's taste in antique furnishings and appointments is a compliment to the period and style of the house.

The flowing roofline, dramatic eave returns, delicate proportions and the deep, rust-red Island sandstone of the Atwell house are a visual treat to passers-by and a continuing reminder to us all of the value of sensible restoration of our architectural heritage. (see p. 60)

Taylor House, North Granville.　　　　- Scott Smith

Strang House, New Dominion.

Taylor House
North Granville

This colourful farmhouse is noteworthy because of its reduced scale — about three-quarters the size of a typical Island ell farmhouse.

It was built around 1895 by Austin Taylor, and after a succession of different owners and tenants, it was acquired in 1983 by Paul and Jean Folland. They have since added a new furnace and flue and a wood stove in the kitchen.

The Taylor House is located on a picturesque site on Route 254 in northern Queen's County.

Strang House
New Dominion

The original centre gable section of the Strang house was built c.1870-1875 by Malcolm MacNeil. The side or ell portion appears to be a later addition, c.1885-1900.

The simple, clean lines of this house have been well preserved. Its discontinuous roofline, eave returns and a somewhat understated front verandah contribute to a restrained and graceful beauty. The present owner is Charlotte Strang.

MacCallum House ("Linden Cottage")
Brackley Beach

MacCallum House, Brackley Beach.   - Scott Smith

The MacCallum house at Brackley Beach on the north shore was built in the early 1820s by Captain James MacCallum, whose father, Duncan, had immigrated to Prince Edward Island in 1770 as a sixteen-year-old from Argyllshire, Scotland. James hired two itinerant Scottish masons to build the house and paid them 75 cents a day for their labour. When Captain James died in 1871, he left the house to his son John.

The early MacCallums were shipbuilders and expert craftsmen. Edward MacCallum, great-grandson of Captain James, operates a lumber and grist mill which is still functioning after more than 160 years of service.

There is a feeling of stability and permanence about both the mill and the house. The stones have been cut with precision and laid with great care, because the walls are still straight and true. These stones were apparently quarried three miles away in the back fields of the farm and hauled to the site on rock sleigh drawn horses.

The original MacCallum house had a wooden-frame rear kitchen wing which was demolished in the early 1950s. It was also in this period that the interior was completely renovated and the large rear shed dormer added. The three frontal shed dormers were built in the 1930s.

The house stayed in the MacCallum family until 1952, when Edward Cutler MacCallum sold it to a Dr. Wortman, who undertook most of the major renovations. Mrs. Wortman sold the house in 1976 to Irwin Jenkins, who did some further renovating of the kitchen. Linden Cottage has been successively owned by Mr. and Mrs. J. Piggot (1979) and Tony and Peggy Sosnkowski (1988), who have converted into a guest house. It stands in a majestic, pastoral setting surrounded by beautiful linden trees that were imported from Scotland. (see pp. 34, 164)

Beaconsfield, Charlottetown.   - Scott Smith

"Beaconsfield"
Charlottetown

This graceful late-Victorian house was built in 1877 by James Peake Jr., the son of shipping magnate James Peake. Unfortunately, the shipbuilding industry was entering a period of decline at the time, and due to a series of financial reverses, the Peakes' occupancy of Beaconsfield lasted only six years. However, it was a period of elite social gatherings during which Charlottetown's aristocracy and international socialites were entertained by the Peakes in grand style.

When William Cundall, who held the mortgage on the house, could not find a suitable buyer, he decided to move in himself. The Cundall family occupied Beaconsfield until 1916, when William's son Henry J. Cundall, a land agent and surveyor, died. A trust was set up to dispose of the house, and for some years it was used as a Y.W.C.A. and nurses' residence for the Prince Edward Island Hospital. In 1972 it was acquired from the trustees as a capital project for Centennial Year, and one year later it became the headquarters of the Prince Edward Island Heritage Foundation, which it is to this day.

Beautifully situated at the foot of Kent Street, overlooking Charlottetown harbour, Beaconsfield replaces an earlier house which was removed from the site in 1876, shortly after James Peake bought the property from J.S. Carvell. W.C. Harris was the architect and John Lewis the builder of this magnificent 25-room

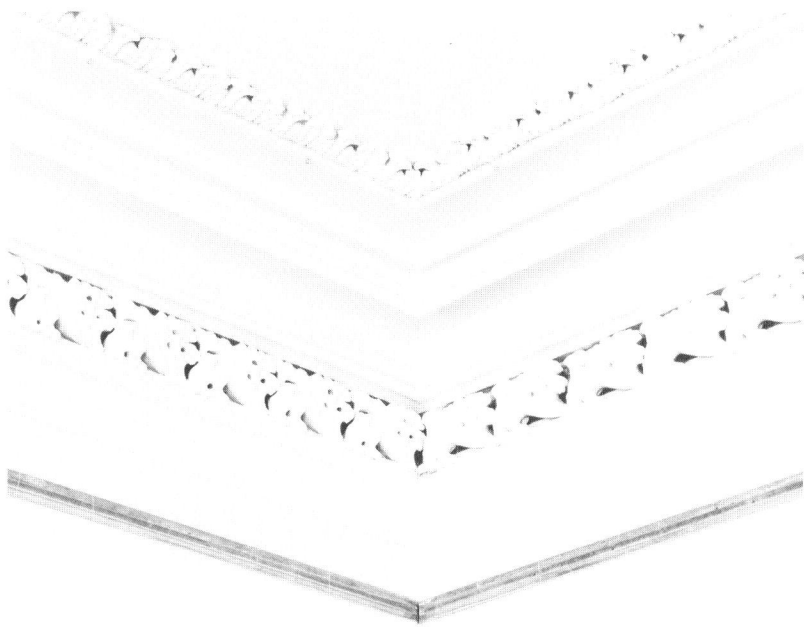

Ceiling mouldings and coloured glass window, Beaconsfield, Charlottetown.

- Lawrence McLagan PAPEI

Second Empire-style house. An elegant belvedere sits atop a mansard roof which contains some beautifully detailed, round-headed third-storey windows, two elements that Harris apparently conceded to the fading Italianate style. A verandah framed by some unusual gingerbread dominates the most prominent aspect of the house, the west elevation, which was originally the back of the building. It is unfortunate that the iron filigree has been removed from the roof and verandah ridges.

Inside, Beaconsfield is luxuriously finished with some marvelously decorative plaster work, ceiling mouldings and marble baseboards. The building was initially heated by nine fireplaces and a few stoves, which have been replaced by a central heating system. The old gas lamps have been removed in favour of new electrical fixtures. An oddity in the basement is a cistern which collected underground water for use in an upstairs "bathing room." A round-headed Romanesque-style window on an interior staircase contains James Peake's initials in coloured and etched glass, a transparent signature of the aristocracy. Original details abound — the chandeliers in the drawing room, the lamp on the newel post (slightly later), the encaustic tile in the front hall, hand-grained woodwork, and bronze hardware on the doors and windows dating to 1877. Beaconsfield is a living museum. (see p. 53)

Appledore, Devon, England. - Scott Smith

Witter-Coombs House, right, can be compared to a typical English row house, left.

Witter-Coombs House
Charlottetown

The Witter-Coombs house on historic Great George Street in Charlottetown is a good example of a successfully restored early row house. Built c.1855 by a family named Witter — who were succeeded in the house, through inter-marriage, by the Coombs — the building exhibits many characteristics of comparative urban models in Britain: the slender, compact proportions, gable-wall chimney and Scottish bay dormer.

Recently, the house was purchased by a local woodworking instructor, Bob Dodderidge, who began an expensive and time-consuming restoration. With the effective use of skylights and an open plan, Mr. Dodderidge succeeded in opening up the 53' x 19' floor plan while retaining the architectural integrity of the orig-inal house. He incorporated the original hand-hewn roof beams into the cathedral ceiling of the third-floor bedrooms and made effective use of the rear two-storey kitchen wing built in the 1930s. The original double entry doors were discovered in the basement, restored and re-installed, and the original casement windows were rebuilt. Mr. Dodderidge refinished the exterior clapboard, trim and shutters, and the colour scheme of mustard yellow and dark brown is most attractive. In 1985 he sold the building to Boyd and Debe Ross.

The Witter-Coombs house is included in a row with four other historic buildings, and during restoration a hidden door was discovered into the old Wellington Hotel immediately next door to the north. (see p. 53)

Lowden House, Charlottetown.   - Lawrence McLagan

Lowden House
Charlottetown

This fine Italianate house on Haviland Street in Charlottetown was designed by architect David Stirling for Esther Lowden, the widow of merchant George Lowden. The Lowdens lost their home on Water Street in the fire of 1866, and shortly after, Mrs. Lowden purchased two lots of what had formerly been barracks land of Fort George but what was later to be known as the prestigious Dundas Esplanade. The house was built in 1868-69 and Mrs. Lowden and members of her family lived there until her death in 1896. The following year the building was leased by U.S. Consul Delmar J. Vail and until shortly after World War I it served as the American Consulate, a necessary establishment considering the substantial shipping trade between the two countries at the turn of the century. Several tenants later, in 1944, it was purchased by the Army-Navy Club, now the United Services Officers Club. The Officers Club is the definitive Italianate house in Charlottetown. The round-headed paired windows, eave bracketing and pediments and the belvedere centred on the low-pitched roof are all trademarks of this squarish Confederation period style. There are also some interesting oval windows in the upper frieze panelling, and the interior features some handsome fireplaces and decorative ceiling mouldings. At one time there were balconies above the front entry and surrounding the belvedere. (see p. 47)

Fairholm, Charlottetown.   - Lawrence McLagan

"Fairholm"
Charlottetown

The architectural distinctiveness of this fine two-storey brick house on Prince Street in Charlottetown amply reflects the social and political status of its builder, the Hon. T.H. Haviland, land agent, Colonial Secretary and Treasurer, M.L.C. and Mayor of Charlottetown from 1857 to 1867. This distinguished politician, whose son was later to become a Father of Confederation, built this grand house in the Georgian style, using Island-made brick. He built it c.1838 with a slate roof, but the front portico was not part of the original house (it was added during the ownership of the Hon. Charles Young). There was initially a conservatory wing attached in which Mr. Haviland grew exotic plants and fruit trees. This annex has been removed, but a vaulted root cellar remains intact. Since the turn of the century an ornate fence and gate have been removed from the property, but in 1929 the enclosed sun porch was built over the front portico. There is also evidence that at one time there

Fairholm, Charlottetown.
- PEIMHF

was a verandah on the south side of the house. Some window sashes have been changed as well.

In 1855 the house was purchased by the Hon. Charles Young, Attorney General, who lived there until his death in 1892. Fairholm was then put up for public auction and sold in 1894 to Benjamin Rogers, a prominent hardware merchant. It has remained in the Rogers family ever since and is presently owned by the Rogers estate.

The formal grace of Fairholm is most evident in its east (front) elevation. The rounded central sun porch, with its decorative fascia panel, is suspended in perfect balance between the two semi-circular bays with their curved glass panes and conical roofs. The stone masonry, highlighted by a string course, foundation and voussoir window lintels, is in excellent condition, and during the summer months the walls are adorned with a sheath of crawling vines. The broad front entry, with its half-panelled sidelights and fanlight, contains coloured glazing which is quite attractive at night. The five blind windows, an architectural technique used to achieve symmetry and balance, are quite rare in Charlottetown, otherwise found only in Province House. The subtle rhythm of its front facade and its beautiful grounds give Fairholm a sedate, picturesque and regal quality. It is perhaps the most outstanding brick house on Prince Edward Island. (see pp. 52, 58)

Belmont, East Royalty.　　　　　　　　- sketch by Robert C. Tuck

"Belmont"
East Royalty

"Belmont" was built in 1810 by the Hon. George Wright, Surveyor-General, on a piece of land that had been Wright property since 1775 and which remained so for another 120 years. Situated in East Royalty, or Bird Island Creek (Wright's Creek, as it was later known), it commanded a magnificent view of the Hillsborough River down the flats to the east.

Georgian architecture is noted for symmetrical facades, refined proportions and elements of neo-Classic architecture. Belmont is typically Georgian under a broad hipped roof with two chimneys at either end of the crest. Neo-Classicism is everywhere, from the wide, panelled corner boards to the pilaster window mouldings and Palladian window on the west facade. The east elevation is dominated by a ground-floor verandah supported by column pairs and a single bay, second-storey balcony that contains an unusual arch within its gabled

roof. This elliptical arch is perhaps attributable to the emerging Federal style and can also be seen in the portico at the west entry. It is unfortunate that some of the original nine-over-six window sashes have been replaced by larger picture windows, but they have been retained on the east side, at least on the second storey. On the west side a fan light is hidden behind the balcony gable, and the first-floor windows have twelve-over-eight sashes.

The building has been well maintained by a succession of different owners since the Wrights sold the property in 1895. In 1975 it was purchased by the Prince Edward Island government from Mr. and Mrs. J.T. Davies for use as office space, but in 1985 it was sold and converted into four apartments and an office for a housing development nearby. It is one of Prince Edward Island's best examples of formal Georgian architecture. (see p. 23)

MacKinnon House, Charlottetown.   - Lawrence McLagan

MacKinnon House
Charlottetown

One of the more sophisticated houses in Charlottetown is this fine brick building at 20 Brighton Road. It was built around 1878 for Donald MacKinnon, who operated a tannery nearby. Soon after, it passed into the hands of a merchant named Frederick Perkins. It served for a period as a Presbyterian manse, and a succession of various owners and tenants followed. Finally, it was purchased in 1970 by Mr. and Mrs. John Clark from a retired Anglican minister, Reverend Thomas Fullerton. The Clarks leased the house until 1973, when they returned from Newfoundland and moved in. Their first task was to build a two-storey addition to the back of the house. They also carried out a series of essential repairs, including the repointing of the brickwork, the installation of a new roof and windows, a new furnace and the reconstruction of some of the pillars in the front portico.

The MacKinnon house is a good example of how prominent exterior elements, designed to interact tastefully, can give a building's exterior rhythm and elegance. In this case the rhythm is provided by the curvilinear elements: the circular portico (a later addition, c.1906), the round balcony columns, the semi-circular balcony projection over the front entrance, and the round-headed dormer windows in the third-floor mansard roof. The well-proportioned mansard, the combination of wrap-around balcony and two-storey front portico, some impressive masonry in the built-up frieze detail, and the sandstone column piers and foundation give this house a stylistic exuberance that is only tempered somewhat by the large and majestic elm trees that surround it.

121

Welsh House
Charlottetown

Welsh House, Charlottetown.   - Lawrence McLagan

Shortly after carpenter Michael Welsh immigrated to Prince Edward Island from Wales and settled in the York area, he built this solid Maritime Vernacular cottage on Cumberland Street in Charlottetown. He built it around 1850, or perhaps a little earlier, using timber cut in York. His initial plan of two separate living units is still retained today. He and his family lived for a long time in the smaller, northern apartment while the larger unit was leased.

Mr. Welsh and his son Michael Jr., himself a fine carpenter and self-taught house designer, finished the house in their own style. The door and window mouldings, elliptical transom light (now on the north facade), stair balustrade, wainscoting, and a wall section of the front vestibule are all handmade and unique. The vestibule, upstairs bedroom dormer, and small office (now a den) were all built by Michael Jr. after the Welshes re-occupied the larger, southern section of the house in the 1930s. He also converted some upstairs space into a bathroom, installed hardwood floors and built a window into the south wall for more downstairs light. Four fireplaces operating from a central flue were sealed when central heating was installed. The two rear kitchen wings, however, were part of the original construction. This is a five-bay house of wonderful scale and character.

Michael Welsh Jr. died in 1947, but his family has continued to live there ever since. His daughter Stella Welsh is the present tenant, and she proudly maintains this colourful and deceptively large house, renting out the smaller unit.

"Birchwood"
Charlottetown

Birchwood, Charlottetown.   - Lawrence McLagan

This beautiful Second Empire-style house was built at the peak of the shipbuilding era on Prince Edward Island by one of the Island's most prominent ship owners and merchants, Lemuel Cambridge Owen. In 1876 Mr. Owen decided that the house he was living in at the time (still standing but in another location) was not substantial enough for a man of his stature. According to the *Charlottetown Examiner* of June 2, 1877, "Mr. Owen himself arranged the plans, Mr. Harper was the builder." William Harper was a local contractor at that time.

Mr. Owen and Mr. Harper certainly captured the essence of the prevalent style of the time, the Second Empire. The central tower and pavilion, mansard roof and wrap-around verandah with its delicate iron filigree roof cresting are unmistakable trademarks of that flamboyant idiom. Unfortunately, the iron crest-ing has been recently removed from the roof and tower crest, but other Second Empire details, such as the paired eave brackets and frieze dentils, have remained intact. A local interpretation appears in the opaque Gothic window heads of the third-floor dormers and the slender pilaster mouldings that frame these windows. Recently, the shingled facades have been superimposed with a layer of more con-temporary siding, but the interior, despite its transition to a funeral parlour, has remained remarkably unchanged. The ceiling mouldings, the old gas chandeliers and mercury door hard-ware were proudly retained by the Aitken fam-ily, who acquired the house in 1895. It became a funeral home in 1963, and a building totally unsympathetic to its design has been built immediately adjacent on Longworth Avenue in Charlottetown.

Peake House, Charlottetown.   - Chris Reardon

Peake House
Charlottetown

A good example of a successful restoration is this elegant brick house on Water Street in Charlottetown. In 1976 it was purchased by the Charlottetown Area Development Corporation (CADC) in anticipation of a waterfront renewal project. In 1978 it was converted into office space and was the home of the Institute of Man and Resources until its transformation into lawyers' offices in 1984.

The Peake House was built c.1836 by James Ellis Peake, one of the Island's leading shipbuilders and a merchant, banker and member of the House of Assembly. He built it in close proximity to his warehouse and wharf and with a panoramic view of Charlottetown harbour. In 1857, in failing health, Mr. Peake auctioned off all his furniture and returned to Devon, England, where he died in 1860. His brother-in-law, T.H. Haviland Sr., lived in the house for a while but later relinquished it to James Peake Jr., who lived there until 1878. After surviving a fire in 1879, the house saw a succession of different owners and tenants — the Leighs, Douses and Batts. By 1976 it had

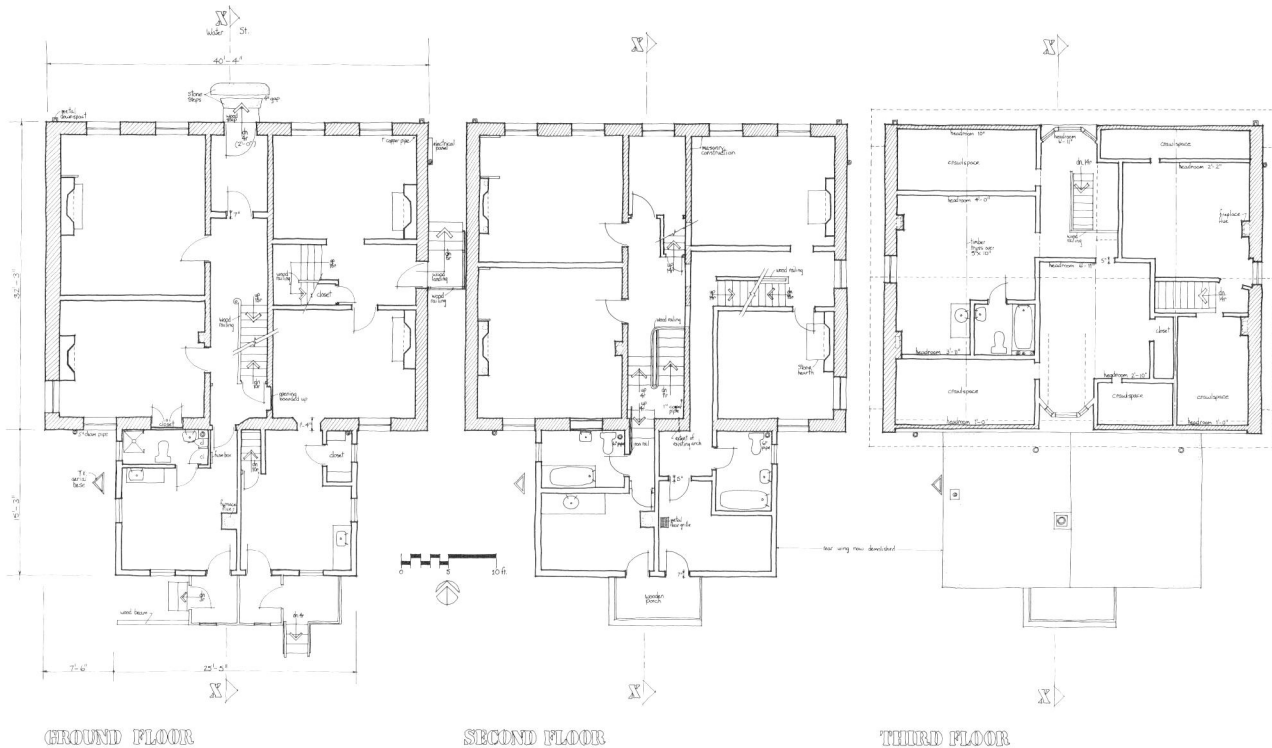

GROUND FLOOR   SECOND FLOOR   THIRD FLOOR

FLOORPLANS: Peake House, Charlottetown, prior to renovation.

become a rooming house and was finally sold to the CADC.

The renovation was extensive and costly. The brick walls and stone foundation were repaired and repointed, and a rear kitchen wing was removed entirely. The reconstruction of chimney flues to make the eight fireplaces operational was a major expense, and the two Scottish bay dormers had to be almost entirely rebuilt. Yet most of the Island-made brick, laid in Flemish bond, and the stone lintels are still in good condition. A new roof was laid with slates imported from Wales. Apart from the removal of some partitions, the interior has not been drastically altered, although the basement has been totally renovated and converted into library space. Some attic space has been claimed for additional offices. The basic centre-hall floor plan has been retained, and a highlight is the fine inner vestibule door with its attractive fanlight transom and trim. Some of the black marble fireplaces have also been restored.

The Peake House is located next to the new courthouse on the Charlottetown waterfront.

Carmichael-MacKieson House, Charlottetown.   - Scott Smith

Carmichael-MacKieson House
Charlottetown

The most notable aspect of the houses of the wealthy and socially prominent is their size. This beautifully detailed house at 238 Pownal Street in Charlottetown is a refreshing exception. Built c.1825 by Colonial Secretary John Edward Carmichael, its modified scale is enhanced by some attractive and well-calculated details, such as the front colonnade, with its slender, paired columns, and the delicate balustrade. The large central dormer, flanked by railings that surround the one-and-a-half-storey house above the eave line, gives an impression of a much larger building. The use of pilasters as window and door mouldings and the implied pediment supported by pilasters in the dormer give the building a mod-ified Classic monumentality. In fact, it is the best example of Palladian neo-Classic domestic architecture on the Island. The clapboard siding, meeting at false quoins at the corners, the intricate, segmental semi-circular transom light above the front door, and the medallion motif in the dormer gable are other details which help to give the house a strong character. One cannot help but notice the large casement windows opening onto the front gallery or the immense twelve-over-eight windows on the north side of the house. Mr. Carmichael obviously placed a premium on having a bright interior. The shallow hipped roof is dominated by two tall chimneys and a small central belvedere.

Carmichael-MacKieson House, Charlottetown. — PAPEI

Robin's Nest, Charlottetown. — Lawrence McLagan

"Robin's Nest"
Charlottetown

After Mr. Carmichael's death in 1828, the house was leased for two years to Chief Justice Edward Jarvis. In 1830 the house and grounds, which contained imported fruit trees and a well with one of the earliest hand pumps in the area, were put up for public auction. The successful bidder was a surgeon, John MacKieson, who for a short period kept an office on the premises. He lived in the house for over 50 years, until his death in 1885, and his diary notes attest to his affection for the building and the great effort he made to maintain it, particularly in the 1860s. MacKieson's descendants eventually sold the house to Mr. R. Sterns, and it has been owned in succession by Dr. and Mrs. George Bishop, Dean Shaw (1986), who undertook a substantial renovation, and the present owner, Reverend Dr. John Foster (1987), who is at present leasing it out.

This delightful Gothic Revival cottage on Brighton Road was built c.1855 by Reverend Dr. L.C. Jenkins, rector of St. Paul's Anglican Church in Charlottetown. He named it "Woodmore," but in 1868 a tenant named Mary M. Robin renamed it "Robin's Nest," a name it carries still.

Remarkably well preserved, the house exhibits many of the Gothic Revival characteristics promoted in the 1850s by A.J. Downing, among them sculpted bargeboards in a tall centre gable and delicate trellis frame work in the wide front verandah. The three-sided bay projection and the half-round window in the front gable, while not Gothic, are prominent and attractive features.

The present owner is Mariedeth Crockett, daughter of Mr. and Mrs. Leonard MacDonald, who moved into this interesting little house in 1937.

McNichol House, Cardigan.

McNichol House
Cardigan

On the outskirts of Cardigan, on a beautiful site overlooking a branch of the Cardigan River, stands a large, rambling, multi-gabled house built partially of Island sandstone. The stone, used only in the first-floor construction, was taken from Casey's quarry at nearby Mitchell River, which re-opened specifically for this event.

The house, which appears to be architect-designed, (possibly by John Hunter), is somewhat eclectic in style, mixing elements of the Chateau and Shingle styles. It was built c.1920 as a summer residence for James McNichol, a United States Senator from Philadelphia, and his wife, the former Margaret Donahue of Cardigan. Unfortunately, Senator McNichol died just prior to its completion. In 1973 it was sold to the renowned Dr. Charles H. Best, co-discoverer of insulin, who also acquired an impressive collection of Maritime antiques and paintings. Dr. Best died in 1978, but his daughter-in-law, Mrs. Eileen Best, has lived there continuously since 1983. There have been few changes to the house over the years except for some redecorating of the interior.

The angularity of the McNichol house, its multi-paned windows, slate roof and handsome Island sandstone make it a singular and attractive addition to the Cardigan landscape.

Nicholson House, Commercial Cross.

Nicholson House
Commercial Cross

This fine example of rural Second Empire-style architecture is located at Commercial Cross, in the heart of the Island's tobacco-growing district in King's County. It was built c.1915 by John Nicholson, a farmer, on a lonely country road once known as the Whim Road, now Route 316. The builder was Murdick Nicholson, and the lumber came from the Cameron farm in Brudenell, the finish wood from Caledonia. The central tower with its fron-tal dormer and mansard roof with delicate trim are trademarks of the Second Empire style. The house also has an attractive colour scheme in beige, green, black and white — a delightful surprise to anyone travelling in the area.

Joseph Vandaele, a Dutch tobacco farmer, bought the house in 1969. He has added a side porch to the house and altered the upper pitch of the roof.

The Goff House, Woodville Mills, following its restoration.

The Goff House, Woodville Mills, in earlier days.   - PEIMHF

Goff House
Woodville Mills

This elegant old estate, known historically as "Cromwell" or "Woodlands," has been rescued from oblivion and successfully restored in recent years by Ron and Mary Cameron, who purchased it in 1976. The architecture of the house is distinctly Irish and reflects the stylistic roots of Goff family builders, who immigrated to the area in the early 1800s. Fade Goff, a native of Wexford County, Ireland, settled at Erindale (or Erin Vale), Prince Edward Island, in 1810. His son John, born in Charlottetown in 1814, went to Launching in 1836 to look after his aging father's estate. In 1840 he settled an estate of 754 acres at nearby Woodville Mills and built the present house in 1841. John Goff was a man of great industry and prominence in King's County. He operated a saw and grist mill on his property for many years and served as a member of the House of Assembly in the 1850s and Legislative Council in the 1860s. He was also the high sheriff of King's County and served a short period as the County's chief magistrate. He died in 1892 at age 78.

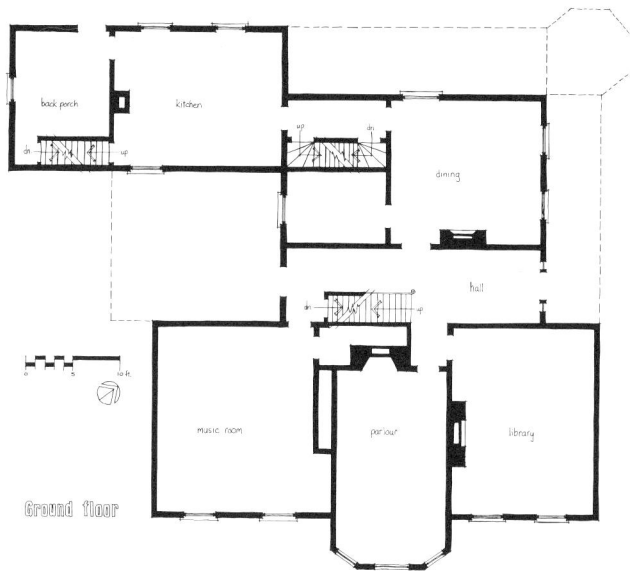

FLOOR PLAN: Goff House, Woodville Mills.

The Goff House is neither austere nor flamboyant in its design. Its charm lies in its clean and simple massing of elements and its graceful proportions. Ornamentation is minimal: pedimented dormer windows with finials and narrow bargeboard scrolls. The lower-storey nine-over-six windows contribute greatly to the proportionate balance of the house.

George Ernest, youngest son of John Goff, inherited the estate, but it was finally sold and the furniture auctioned off by George's son Edward around 1955. A succession of different owners followed. It was used briefly as a bunkhouse in a logging operation and as a schoolteacher's residence, and in 1967 it was bought by Victor Oland, a former Lieutenant-Governor of Nova Scotia who had planned to restore the old estate. Unfortunately this never happened,

and despite some essential maintenance, the house slipped into a period of decay and vandalism until its rescue in 1976 by the Camerons.

The restoration has been a long and expensive one. The sandstone foundation required a major reconstruction, and the interior walls have been almost completely rebuilt. A wood furnace and a backup oil furnace have been installed in the basement, a far cry from the open-hearth fireplace in the kitchen that, as recently as 1955, supplied not only a cooking facility but heated a section of the house as well. The Camerons plan to restore this fireplace at a later date. At present they have completed the reconstruction of two of the four brick chimneys in the building. Verandahs on the north and west sides of the house and a breakfast room on the south side were all removed prior to the Camerons acquiring the property.

One of the most impressive aspects of the Goff estate is its grounds. The property is surrounded by a stone herringbone fence, rarely seen on Prince Edward Island. This fence, threatened with destruction at times by road builders, is still in good condition, and it encloses a property that is enhanced by a variety of linden, elm, oak and sugar maple trees, many of which were imported as seedlings from England and Ireland in 1845. Behind the house are outbuildings whose neat and simple lines are in sympathy with the architectural character of the main house. It is heartening to see this noble old homestead being so tastefully restored when many like it have fallen victim to time and disuse. (see p. 58)

(see p. 58)

Aitken House, Lower Montague.

Aitken House
Lower Montague

One of the most charming and historic of the Island's sandstone cottages is the Aitken House in Lower Montague. Its design and sturdy construction draw heavily on the Scottish origins of its builder, George Aitken.

The house history has been extremely well documented, and the fact that it has always remained in the Aitken family has simplified the researcher's task considerably.

John Aitken, a native of Auchenhay, Kircudbrightshire, Scotland, came to the Island of St. John's with his wife and four children aboard the *Lovely Nelly* in 1775. For $60 he purchased 100 acres of land on the Montague River oppo-site Brudenell Point where de Roma's fishing and trading settlement had been. No trace remains of the original log house that he built there. It was his son George, born in January 1779, who built the stone house that now stands proudly overlooking the Montague River. In 1815 George bought 90 acres of land adjacent to his father's property and erected a grist mill. The profits from this mill enabled George to consider building a house, and after a short trip to his grandfather's home in Auchenhay, Scotland, he returned with plans for a new nine-room, one-and-a-half-storey cottage, 38' x 24'.

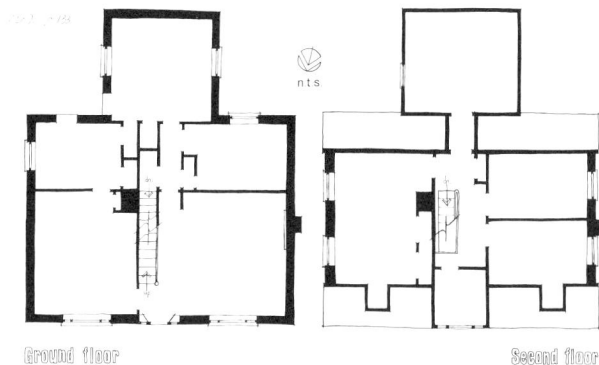

FLOOR PLAN: Aitken House, Lower Montague.

Six-over six window, Aitken House, Lower Montague.

Construction began in 1843, and in 1848 George and his wife Hannah celebrated their golden wedding anniversary in the new house. It took the first year to quarry and dress enough stone to build the cellar and first-floor walls, which are 36 inches thick. The gable-end walls both contain built-in chimney flues; however, there is no evidence of any fireplaces except in the rear kitchen wing. Huge juniper beams were used in the construction of the ground floor and oak was also used extensively in the interior woodwork. Wooden window lintels were used instead of stone, which was not available in that length. The front entry hall is lit by two half-sidelights that frame a "Christian" door, so named because its panels form the shape of a cross. The central dormer directly above and the two front rooms all contain large, tri-sectional, six-pane windows, some of which still contain original glass. One unique feature of this house is a small recess in the wall beside the hallway window which has continually housed a telescope for scanning the river and arrivals at the house.

The Aitken house is perhaps the most sound, both structurally and thermally, of all the Island's stone houses. In a letter written to the August 13, 1888, edition of the *Daily Examiner*, Island architect W.C. Harris reported that "the stones in the walls were perfect, still retaining the tool marks made when they were cut half a century ago. The inmates said that the building was always comfortably warm in winter and never damp."

Fred Aitken is the present owner of the house, and although some changes have been made to the internal planning, an impression of a living museum remains. The house is alive with antiques such as George Aitken's old family trunk, studded and leather bound, which at one time held all the important documents of business. Fourteen hand-turned and wooden-pegged chairs are still in use, and some salt-glaze stoneware can be found in the kitchen. History abounds within George Aitken's remarkable house. (see p. 58)

Tweedy House, Vernon River. — Scott Smith

Smith House, New Perth. — Scott Smith

Tweedy House
Vernon River

The Tweedy House was built around 1840 by Thomas Tweedy, whose family had immigrated to Prince Edward Island from Yorkshire, England, ten years earlier. It has remained in the Tweedy family ever since and is presently owned by Mrs. Irving Tweedy, whose late husband was a grandson of Thomas.

The house itself is a typical Maritime Vernacular cottage with a gabled front dormer, but the continuous eave and the elegant sidelights and transom light around the front door give it some distinction.

Smith House
New Perth

The mansard roof is quite common in the eastern part of Prince Edward Island, and this fine house at New Perth is typical of that Second Empire style. Its steep green mansard and symmetric front facade stand out among several other houses of similar design in the New Perth area of King's County on Route 3.

The house was built by a family named Stewart and purchased by Alexander Smith in 1854. Around the turn of the century he built the mansard onto what was originally a flat roof. His great-grandson Elliot is the present owner of house and farm.

Octagonal house, New Perth.

Octagonal House
New Perth

Jack Williams' converted octagonal barn was hauled about one mile across the fields on skids from a farm on the Baldwin's Road. It was built by Micipsa Moar as a three-storey-and-basement forge, workshop and machine shed. Mr. Moar was a highly skilled wheelwright of Scottish ancestry who specialized in wagons, sleighs, coffins and cabinetry. The barn was vacant for many years before Mr. Williams bought it in 1967 and had it moved to its present location. It sits at the edge of a lovely pond, also created by Mr. Williams, on Route 3 in New Perth. The conversion into a house, carried out by Mr. Williams with the help of carpenter Gordon Reid, has generated some interesting interior spaces. A focal point is a sandstone fireplace built from an old outdoor oven found at St. Mary's Bay. A winding iron staircase leads to the second floor. An addition housing a furnace and bathroom was built soon after the barn was relocated. The house is estimated to have been built around 1885 and measures 32 feet in diameter.

Quinn House, Cardross.

Quinn House
Cardross

A good example of a vernacular house with a well-defined ethnic origin is the Quinn house in Cardross, King's County. James Quinn, an immigrant from New Ross, Wexford County, Ireland, settled on his 100-acre farm around 1850. James, his wife and six children lived temporarily in a simple log cabin, and it wasn't until about 1865 that they built the present house, set deeper on the property from the road. The design of the house — steep gable roof, low eave line, symmetrical plan with kitchen wing addition and minimal fenestration — draws heavily on the austere, functional, small farmhouse built in Ireland in the late eighteenth and early nineteenth centuries. The only apparent concessions to a decorative style

are the sidelights and transom light that surround the front door and the vertical wainscoting on some interior walls. Construction is quite substantial, as seen in the exposed hand-hewn beams in the kitchen and some wide pine board sheathing exposed on one wall. The colour scheme is white with green trim.

Mary Quinn, great-granddaughter of James Quinn, is now the owner of the house and lives in the kitchen wing in wintertime. She remembers when there was an open fireplace in the kitchen. There have been few other changes. Electric power was installed about 1960 and the house has been insulated. Indoor plumbing was not installed until 1982.

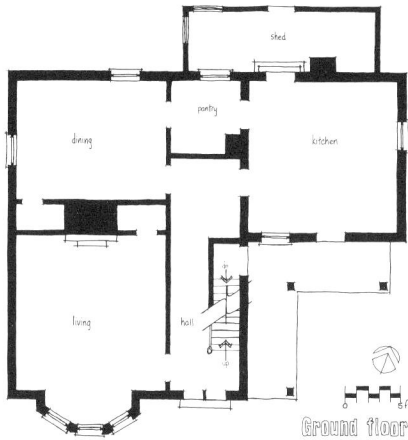

FLOOR PLAN: Bourke House, Millview

Bourke House
Millview

Bourke House, Millview.   - Scott Smith

This house, located on Route 3 in Millview, King's County, is one of the most stylistically significant houses in Prince Edward Island. It embraces in a most harmonious way two distinct design trends: the steep mansard roof of the Second Empire style and the traditional ell farmhouse with ornamental window hoods of the Gothic Revival. With its bold colour scheme of gold with brown trim, seldom seen outside areas of Acadian settlement, this house creates a very strong visual impression.

The building was begun in Cherry Valley by Jack Jones about 1880 and shortly thereafter was purchased by a magistrate named John Roach Bourke, who hauled it to its present location and completed its construction. After the residency of two more generations of Bourkes, the house was sold to a family named MacLeod in 1927. The MacLeods dismantled a rear kitchen wing, removed it to a farm down the road, and replaced it with the present back porch. About thirteen years later it was purchased by Robert Drake, who proudly maintained this most characteristic house until his death in 1980. Originally white with green trim, the house was repainted by Mr. Drake, who also removed the iron roof cresting, also characteristic of the Second Empire style. The highlights of the interior are the handsome marble fireplaces, still operational, and the upstairs walls which slope inward, reflecting the outer mansard roofline.

Barry and Shirlee Hogan, who bought the house in 1986, have been quite busy repairing the foundation, improving the landscaping and repainting the entire building.

Judson House, Alexandra. - Regan Paquet

Judson House
Alexandra

When George Judson, a skilled carpenter and farmer, decided to build a house for his large family, he sent away to Pennsylvania for a set of catalogue house plans that were quite accessible around the turn of the century. In 1897, after building a model to see how it might turn out, he built this Queen Anne-style house in Alexandra, a small farming community of essentially English settlement on Prince Edward Island's south shore. Its tall, beautifully proportioned corner tower overlooks the long, low flats that run down to the shore of Northumberland Strait. In 1986 Beaton's Farms purchased this house from the estate of Bruce Judson, whose family, six generations earlier, in 1793, held one of the first freehold leases in the area. (see also p. 30)

Howe House, Guernsey Cove.

Howe House
(Strait View Farm)
Guernsey Cove

The Howe House in Guernsey Cove is another house which has remained in the same family through four generations. Built by William Howe, an English shipwright, it is presently owned by his great-grandson Maurice Howe. William bought the property in 1842 from John Duffus, but his first house is now one of several barns on the farm. The present house was not built until about 1875, and the tall central gable and beautiful detailing have remained untouched over the years.

The original pine shingles still clad the walls, but the roof had to be reshingled in 1963, revealing large sheets of birchbark insulation beneath it. In 1931 the high concave kitchen ceiling was replaced by a lower flat one. Apparently, William had an asthma condition and felt that a high ceiling would improve air circulation. A rear porch has been rebuilt twice, in 1916 by Maurice's father and in 1976 by Maurice himself.

William Howe built a very sturdy house with hemlock sills a foot wide on a sandstone foun-dation and pine flooring two inches thick. He said that he did not want a house that he would "have to go out and prop up when the wind blew!" He also built a house with a fair degree of neo-Classical charm. The broken pediment over the front door is a detail frequently seen in the Guernsey Cove area, and the shelf-headed windows complement it nicely. The four-panel front door, which used to be protected by heavy storm doors, is surrounded by sidelights and a transom light and surmounted by an attractive triparite, triangular-headed window. The dentil frieze decoration beneath the eaves and the upper window hood moulding give depth and dramatic impact to the front elevation.

The Howes are justifiably proud of their house and have recently undertaken a renovation of the interior. At one time a long hawthorn hedge row ran 400 feet along the road frontage, and extensive flower beds gave this neat and stylish house a reputation as "the showplace of the Cove."

Beck House, Guernsey Cove.

Beck House
Guernsey Cove

Vere Beck, an engraver, and his wife, Elizabeth Sarah, emigrated from Crayford, Kent County, England, to Prince Edward Island in 1813. All the Becks on the Island are descended from their twelve children. In 1814 Vere purchased 100 acres of land in Guernsey Cove from John Cambridge, proprietor of Lot 64, and built his first log cabin there. Although Vere promptly sold 50 acres on the east side to a man named Henry Brehaut, the remaining property has stayed in Beck hands for four generations, the present owner being Windsor Beck, Vere's great-grandson. Vere died in 1878 at the age of 95, after a long and productive life which included service in the Legislative Council from 1837-43. His old family Bible, brought from England, is still at Windsor Beck's house, as is a wooden-pegged boot, the first shoe ever worn by Bartholomew Beck, Windsor's father, and handmade by William Beck, Windsor's grandfather.

Windsor Beck's house is the fourth to be built on the farm. Apart from the original log cabin built much closer to the shore, two other houses that were built now serve as outbuildings. The present house was built about 1870 by William Beck, and with Windsor now in a nearby nursing home, it is deteriorating rapidly.

The house is simply built in the Maritime Vernacular style, with a gabled dormer atop the front door. Decoration is restricted to eave returns and sidelights and a fan light over the front door. One notices the absence of power lines leading into the building. This is because the Becks have never felt the need for electric power. Apart from reshingling the roof, patching the sandstone foundation and installing a rear staircase and an archway to the dining room, the Becks have made no major changes to the house. Birchbark insulation was found in the roof structure during reshingling.

Dr. Roddie MacDonald.
- courtesy Jean & Colin MacDonald

MacDonald House, St. Peter's.

MacDonald House
St. Peter's

Dr. Roderick MacDonald, a long-time family physician in St. Peter's, acquired this property on the south side of St. Peter's Bay in 1891 from John Nichols. The original house, the residence of a local merchant, William Ernest Scott, had only three rooms and a flat roof. This early house, probably built about 1880, was apparently moved to the present site from another location, and in 1892 Dr. MacDonald began its enlargement and embellishment. He built the bellcast mansard roof and added the corner tower and front verandah with its decorative gingerbread. He also added a rear wing, probably to accommodate his seven children who were born in the house. Two of his offspring, Colin and Jean, are the present owners, having inherited the property upon their father's death in 1961 at age 103. Dr. MacDonald was a well-respected and dedicated family doctor who travelled far afield to make house calls well into his nineties. This fine house, on Route 2 overlooking St. Peter's Bay, is a fitting monument to this remarkable man.

Clow House, Murray Harbour North.

Clow House
Murray Harbour North

The Clow House is an excellent example of a building whose character has been drastically altered by secondary embellishments. The house was built c.1865 by James Clow, the son of Scottish immigrant Benjamin Clow. James was a prosperous shipbuilder, lobster packer and merchant. His son Benjamin took over this flourishing mercantile business in 1890, and it was he who hired some skilled carpenters from Sturgeon to embellish the house. In 1909 Alex Finlay and his son Russell built the two corner bays with conical roofs,

the interconnecting balcony with its lovely balustrade and gingerbread, and the central mansard-roofed turret. The original house was just a plain box, much the same as the side wing, which unfortunately has been left as it was built.

Benjamin's business went bankrupt in 1928, but his two sons, James and George, operated a provisions store in a nearby converted warehouse. The brothers, both retired, are the present occupants of the wing and the house. (see p. 164)

(see p. 164)

Prowse House, Murray Harbour.

Prowse House, Murray Harbour.   - PAPEI

Prowse House ("Riverside")
Murray Harbour

This brightly coloured mansard-roofed house has belonged to three successive generations of Prowses since its construction in the early 1880s. It was built by an English carpenter named Richard Murley for Senator Samuel Prowse. Samuel's father, William Prowse, was a farmer who immigrated to Prince Edward Island from Devonshire, England, in 1830. Despite these agricultural beginnings, Samuel became involved in the processing and exporting of fish, lobsters and produce. He also served in the Provincial Legislature in the 1890s. His flourishing business in Murray Harbour South was eventually taken over by his two sons, William and Albert. In 1902 Samuel died, and William sold his share of the business to Albert, who continued to operate "Prowse and Sons"

on his own. Albert, who succeeded his father in the house, also became involved in politics and was elected to the Prince Edward Island Legislature in 1899 and 1904. One of Albert's ten children, his son Gerald, operated a hotel in the house between 1958 and 1965. The present owner is his widow, Mrs. Constance Prowse.

Very few changes have been made to "Riverside," as it has been known, although there was once a railing and a balcony atop the front verandah. This beautifully detailed verandah and the bright blue clapboard siding make the Prowse house the most visually prominent building in Murray Harbour South. It is located just at the northern end of the bridge spanning the South River.

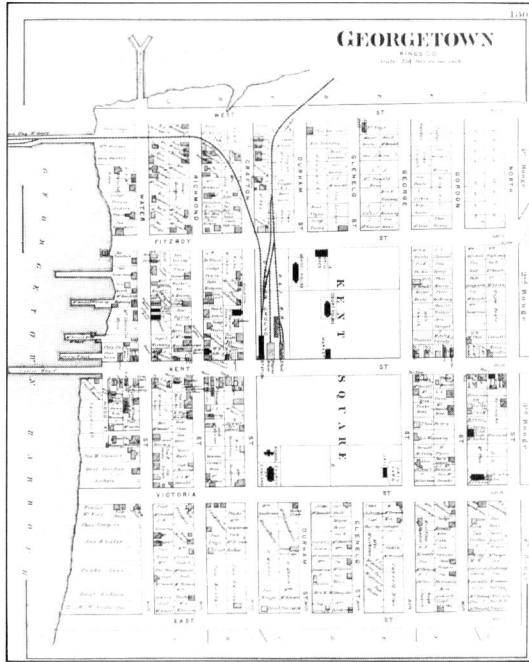

- from *Meacham's Atlas*

Georgetown

In 1765 Captain Samuel Holland named Georgetown the County Town of King's County, primarily because of its excellent harbour. Its population grew from a low of 37 in 1832 to 1,250 in 1871, due to a tremendous boom in shipbuilding and the import/export trade. Some say that the arrival of the railroad in 1872 precipitated a long and painful post-Victorian decline in the town's fortunes. Today its population stands at about 750, but its shipbuilding, lumbering and fish-processing industries are showing signs of recovery and stability. Georgetown also boasts a substantial inventory of heritage buildings.

"The Highlands"
Georgetown

In a grove of spruce trees just west of Georgetown, King's County, stands one of the Island's finest examples of Queen Anne Revival-style architecture.

"The Highlands" was built in 1893 as a summer house for Donald Alexander MacKinnon, a native of Uigg, Prince Edward Island, and Lieutenant-Governor of the province from 1904 to 1910. A graduate of Dalhousie University law school, he practised law in Charlottetown until 1901, when he was elected to the House of Commons representing Queen's County. He was a man of imposing stature and was described by one of his colleagues as a "constructive statesman." He died in Charlottetown on April 20, 1928, at the age of 65 years.

In the 1930s and early 1940s the Highlands was operated as a hotel by George MacKinnon of Sherbrooke, Quebec. It was in this period that the adjacent Pavilion, a dance hall, and twelve motel units, some of the first in Prince Edward Island, were built. Built in the Shingle style, the Pavilion contains an impressive fireplace and chimney constructed from 90 tons of South American ballast rock.

Donald Alexander MacKinnon.
 - PEIMHF

The Highlands, Georgetown.

The Highlands Hotel had its own golf course and private railway station, and the weekly summertime dances in the Pavilion, with live orchestra, are remembered with fondness by some local residents, including the present owners, Mr. and Mrs. Wallace Rodd of Charlottetown. The rear section of the ground floor (see floor plan p. 146) contained the hotel kitchen and the old ice refrigerator. Monogrammed cutlery and dishes remain. A pamphlet advertising the hotel proclaims that "even the most blasé culinary connoisseur will take no exception to the fresh fish just out of the water . . ." Inclusive rates at that time were $25 to $35 per week! Upstairs in the main house nine bedrooms contain some of the pine dressers and old beds from the hotel days. There is still a sink in each of the rooms as well.

The hotel was operated until the advent of World War II, when it was put on the market through a trust company and eventually bought by Reverend Dr. John Sutherland Bonnell, an

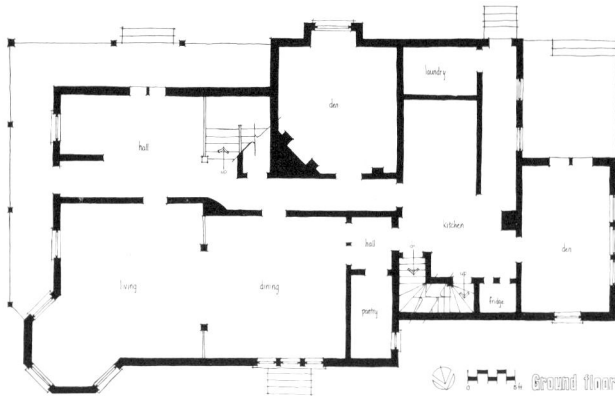

FLOOR PLAN: The Highlands, Georgetown

The Pavilion, The Highlands, Georgetown.

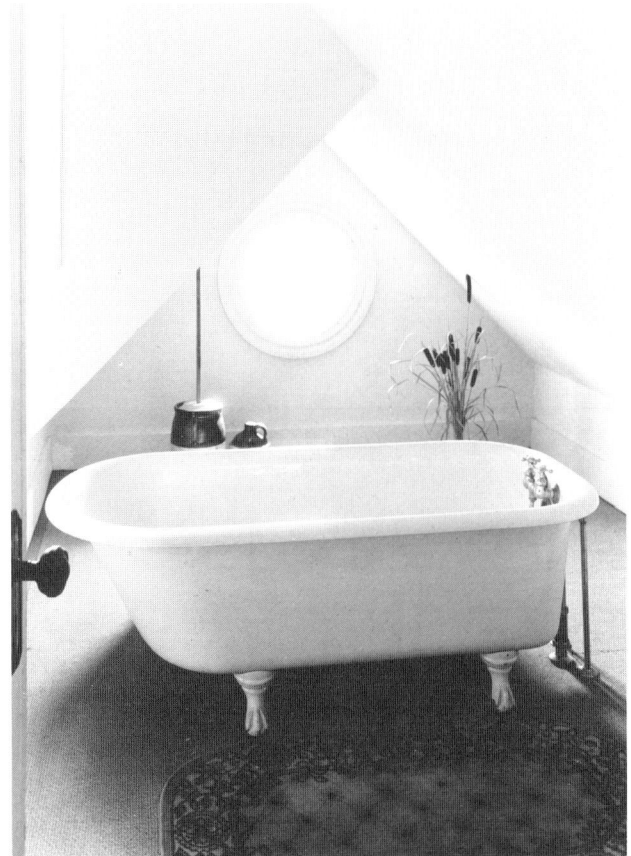

Bathroom, The Highlands, Georgetown.

Island native and long-time minister of Riverside Presbyterian Church in New York, who used it for about 35 years as a summer home. It was then bought by Mr. and Mrs. Rodd in 1971, who continue to use it as a summer home.

Beautifully sited and with a tasteful and sympathetic colour scheme, the Highlands is well-preserved and exhibits all the major charac-

teristics of the Queen Anne Revival style (1880-1900). A polygonal corner turret with tent roof, irregular massing reflected in the multi-gabled roof line, an encircling front porch, alternating horizontal clapboard siding and fish-scale shingles, and a third-floor variant of the Palladian window all combine in an extremely harmonious way. (see also p. 49)

146

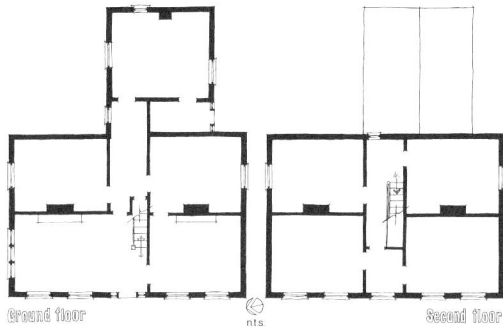

FLOOR PLAN: Fairchild House, Georgetown

Fairchild House, Georgetown.

Fairchild House
Georgetown

On the very edge of Georgetown harbour and protected by a retaining wall of granite ballast stone from Scotland, stands the Fairchild house, a well-preserved example of Classical Revival architecture, New England style. Its history belongs to the sea, since it was built c.1845 by a boatbuilder, Joseph Fairchild, and inherited by his son Nelson, a licensed deep-sea captain who ran the first ferry between Georgetown and Lower Montague. This paddlewheel boat was also built by Joseph, who specialized in barques designed for commercial shipping.

The house has always been in Fairchild hands. Nelson's son Joseph lived in the house briefly, succeeded by his two daughters, Kathleen Fairchild Way and Rose Fairchild Halpin. Kathleen and Rose were both born in the house and have successively used the house only as a summer residence, living the rest of the year in the southern United States or in Charlottetown.

The Fairchild house is very solid and built to last. Its straight, clean lines, gabled roof and symmetrical front elevation attract the eye that scans the historic Georgetown waterfront. Over the years there have been some changes. A porch that used to project beyond the front door, with its attractive sidelights and transom window, has been removed. A rear kitchen has been converted into a den, and the Nova Scotia freestone foundation has required brick infilling. A new brick fireplace has been built in the parlour, replacing the old coal stove that used to supply the heat. The original pine floors with their square nails and the old marble fireplace in the dining room are still intact. The Fairchild house stands as a proud reminder of Georgetown's illustrious past.

Lavandier House, Georgetown.

Lavandier House
Georgetown

The Lavandier house in Georgetown was originally a section of William Sanderson's warehouse, which was located a little farther east on
Grafton Street in Georgetown. Sanderson, a
native of Scotland, was one of Georgetown's
earliest provision merchants, having settled in the
area in 1832, when the population of Georgetown was a mere 27. He established his business
around 1840, and it was later operated by James
Easton. Later still, the house was separated from
the store and moved 200 yards west on Grafton

Street. F.W. Lavandier bought it in 1944 from Ted
Easton. The Lavandier family emigrated from
France and are distantly related to Jean-Pierre de
Roma, founder of one of the Island's earliest
French settlements at Three-Rivers in 1732.

The house's black-and-white colour scheme
has a vivid and dramatic impact. The original
cupola is still intact, but Mr. Lavandier has added
a front porch and shed, changed the windows,
added shutters and done some commendable
landscaping.

MacDonald-Mair House, Georgetown.

MacDonald-Mair House
Georgetown

Malcolm MacDonald was a prominent Georgetown merchant, of the MacDonald-Westaway Company, and a descendant of Andrew MacDonald, patriarch of the MacDonald family in Prince Edward Island. His house on the corner of Water and Kent streets in Georgetown has a colourful past. The present owner, Tim Mair, believes that it was built as early as the 1840s as a warehouse on the waterfront and that it was moved to its present location a few years later, possibly in 1851. There is evidence to support his claim. In the basement, remnants of the framing of an old freight elevator can be found. In 1890 the house was moved again, to a deeper position on the present lot, and it was at this time that it received most of its embellishment: the twin frontal bays whose roof structures blend nicely into the sim-

ple gabled roofline, the delicate trim and windows with unique upper sashes. Concurrently, the roof pitch was steepened and new fireplaces were added.

Mr. Mair, who purchased the house in 1967 from Dr. Alec Kennedy, has gone to great lengths to preserve the architectural integrity of the building. The original hand-split shingles he removed from the exterior now sheath the interior walls of a small kitchen addition built onto the rear of the house. He has also rebuilt a railing over the front porch based on an old photograph of the house. The deeply grained interior doors by Mr. William Crossman from Southport have been painstakingly restored and the richly coloured facades are a landmark at this prominent Georgetown intersection. (see p. 50)

Underhay House
Eglington (Bay Fortune)

Underhay House, Eglington.

In rural Prince Edward Island, buildings constructed of Island-made brick are very rare. The Underhay house in Eglington, King's County, is one such survivor, having been built entirely of brick shipped up the coast by schooner from Southport. A mason by the name of Pratt from St. Peter's built the house in 1876 for John Collier Underhay, a successful farmer and member of the Legislative Assembly in the late 1800s. His father, William Underhay, a former sheriff and magistrate, had emigrated from Devonshire, England, and located on an adjacent farm in 1818. The house has remained in the Underhay family ever since. It has been occupied in succession by John Colliers' son William Henry and his grandson, Reid, who took possession of the house in 1926. Both of these men were prosperous farmers, and Reid's son John has continued this tradition. The farm is now known as "Ayrslie Farm."

Over the years the house has undergone some minor changes, but the deep, truncated hip roof, recently reshingled, has remained intact, as has the gabled kitchen wing in the rear. The original shed dormers have been changed to the present gabled ones, and the existing louvred shutters were initially one-piece storm shutters. The centre-hall floor plan has been maintained, although an interior partition has been removed, creating a full-length living room on the ground floor. With the advent of indoor plumbing, one of the upstairs bedrooms was converted into a bathroom.

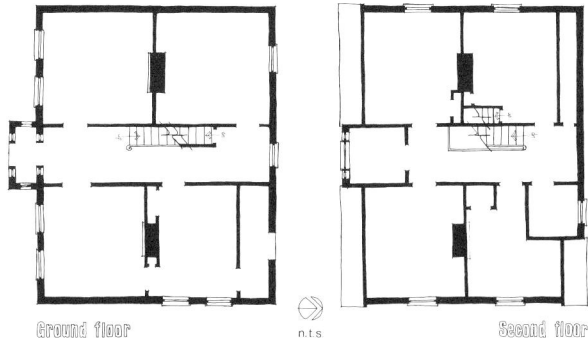

FLOOR PLAN: Beaton House, Souris.

Ground floor   n.t.s.   Second floor

Beaton House
Souris

Beaton House, Souris.

This delightful, one-and-a-half-storey Maritime Vernacular house in Souris has a long and interesting history. It was built about 1850 by Donald Beaton, a prominent fish merchant and MHA, whose family was among the earliest settlers at nearby East Point. Research indicates that Mr. Beaton built the house just prior to his marriage in 1854, but there is evidence that a summer kitchen wing removed by the most recent owner, the late Raymond A. Leard, may have been an earlier house. Donald Beaton died in 1865 of diabetes, and his widow, Clementine, sold the house to Caleb Carleton Sr., a lobster packer and United States consul agent in Souris who was in charge of clearing American fishing boats. His son Caleb Jr. took possession of the house after his father's death in 1904, and it remained Carleton property until 1953, when it was sold to Mr. Leard, himself a well-known merchant and historian. Mr. Leard's widow and son continue to live in the house.

Ray Leard did a lot of work on the house. He reshingled the roof and built an enclosed front porch whose New England-style entry is a tasteful and harmonious complement to the other neo-Classic elements of the front facade: a wonderful Palladian window in the upper-storey dormer, corner boards and eave returns, and nine-over-six, shuttered windows with pilaster side mouldings and entablature heads. This beautifully composed front elevation exhibits many characteristics of the New England neo-Classic style, probably promoted by visiting American sea captains, but nonetheless superimposed on an essentially Maritime Vernacular style. Mr. Leard provided bathroom space on the second floor by adding a rear dormer, and he replaced the old floorboards with a hardwood floor throughout.

Driving through Souris, one cannot help but notice this historic house on the north side of Main Street overlooking Colville (Souris) Bay.

151

McIntyre House, Souris.

McIntyre House
Souris

The tremendous success of the shipping, fishing and mercantile trades in Souris in the late nineteenth and early twentieth centuries generated a number of substantial private homes. Many were built in the Second Empire style, and indeed there seemed to have been a modest proliferation of that distinctive mansard-roofed trend right across King's County. In the Maritime Provinces generally, the Second Empire style was interpreted in wood-frame construction in a picturesque and ornamental way. There was a wealth of surface detail in the window surrounds and eave bracketing, and one of the most regionally distinctive features was the three-sided or bay dormer, often found with round-headed windows. These particular elements and an extremely deep mansard roof dominate the front elevation of the McIntyre House on High Street in Souris East. It was built about 1875 by Sandy Clark, the

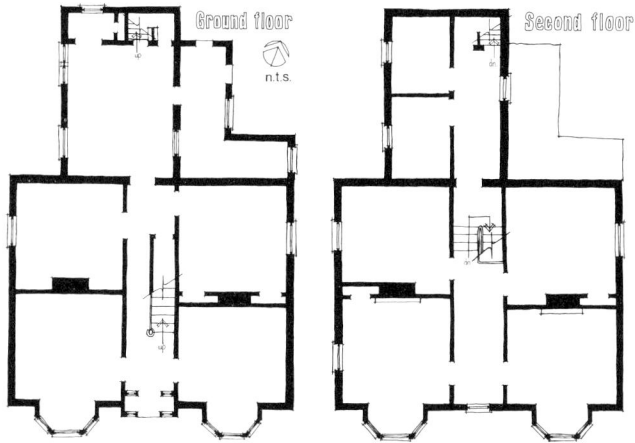

FLOOR PLAN: McIntyre House, Souris.

Shaw House, Souris.

local rail station agent, for Dr. P.A. McIntyre, a physician who was later (1899-1904) to become the Lieutenant-Governor of Prince Edward Island. The house passed through a number of hands, including J.J. Hughes', before Arthur Peters, a prominent fishbuyer, bought it in 1940. There is evidence that during this period the upper roof slope was added and the original conical tops to the front bay projections were reduced to their present truncated profiles. The original back porch is still intact, however, and the interior features a lovely central spiral oak staircase. Mr. Peters (originally an Acadian name, "Pitre") lived in the house until 1979-1980, when it was converted into a group home for mentally handicapped adults. It was substantially renovated at that time.

The McIntyre house is impressive as a building of great mass, but its significance as one of the Island's best examples of Second Empire architecture cannot be overlooked.

Shaw House
Souris

This peculiar Italianate house was built in 1894 on a triangular lot directly across the street from its present location. It was built by Bernard Creamer to a design by W.R. Dingwell — an unusually stylized response to an awkward site. L. Eustace Shaw, who successfully operated a jewelry store on the premises, was the original owner. For many years Dr. A.A. (Gus) MacDonald made the house his home and office, and in 1947 Frank Case moved the building across the street to its current site. A rear summer kitchen was discarded in the move. Since 1972 the house has been owned by Mr. and Mrs. James Hughes.

MacLaren House, Belle River.

MacLaren House
Belle River

This colourful ell farmhouse located in Belle River was built c.1885 by "Red Hector" Mac-Millan, a skilled and imaginative carpenter from nearby Wood Islands. He built it for Daniel MacLaren, a local shopkeeper, whose son Will built the adjacent farm and outbuildings. Will inherited the house upon his father's death and lived there with his wife until 1940. At that time, Miss Priscilla Bell and her widowed mother were just returning to the area after living in the United States. Miss Bell, along with her nephew and his wife, have been living there ever since.

Repainted in 1987, the white house with green trim and its contrasting dark red roofing shingles has an appealing richness. There remains ample evidence of Red Hector's skill in the vertical hardwood wainscoting and the unique verandah framing. The house has five bedrooms on the second floor, and apart from the conversion of a small downstairs bedroom into a bathroom and the installation of indoor plumbing and electricity, there have been few changes. The kitchen, pantry and dining areas are in the ell wing, and the living areas are located in the main body of the house.

Murchison House Holm Cove.

Murchison House
Holm Cove

This well-preserved house is one of the earliest on the Island's south shore. It was built c.1823 by Samuel Murchison, a Selkirk settler, whose son married the daughter of Dr. Angus McAulay, agent for Lord Selkirk.

Initially, the house stood closer to the shoreline on the 120-acre farm to facilitate access by boat, but it was moved closer to the road in 1908. At this time a lateral kitchen wing was added to the house. In the mid-1950s a second rear wing (now an outbuilding) was removed. The main body of the house still retains the clean, time-tested symmetry of its Scottish ancestors.

The house and property remained continuously in Murchison hands until 1957. The present owner is Elizabeth Townsend of Halifax, Nova Scotia.

Rectory, Presbytery and Manse

see also frontispiece, p. 57

Detail from presbytery, St. John the Baptist Roman Catholic Church, Miscouche (c.1891).

The Sacred Heart Roman Catholic Church rectory, Alberton (c.1894), designed by Summerside architect George Baker and built by George Gard.

St. Patrick's Roman Catholic Church rectory, Ft. Augustus (c.1875), quite possibly designed by architect Thomas Alley.

Presbytery, St. Simon and St. Jude Roman Catholic Church, Tignish (1872). - Scott Smith

This rectory of Notre Dame de Mont Carmel Roman Catholic Church was designed by architect R.P. Lemay in 1902. - Scott Smith

Presbytery, St. Mary's Roman Catholic Church, Souris.

Details

Entrance, Richard Clarke House, c.1864, Orwell Corner.

Bay windows, Pinette.

Balcony detail, Sharpe House, Summerside.   - Lawrence McLagan

Bellcast mansard roofline, Pigott House, Savage Harbour.
- Scott Smith

Eyebrow dormer, Furness House, Vernon River.   - Scott Smith

Attic window, Vernon River.

Gingerbread, MacKinnon House, Freetown.
- Regan Paquet

Entrance detail, Spring Street, Summerside.
- Scott Smith

Dormer, Morrison House, Summerside.
- Lawrence McLagan

Dormers, Pollard House, Cape Traverse.   - Scott Smith

Dormers, St. Peter's Road, Charlottetown.   - Scott Smith

Detail of gateposts, Sir Andrew MacPhail House, Orwell Corner.

Interior doorway, Wyand House, Mayfield.

An interesting detail from the MacCallum House in Brackley Beach where the MacCallums and successive families sharpened their knives.

Balcony gingerbread, Clow House, Murray Harbour North.

Detail from Penny House, Beach Point.

Dormer detail, John Charles MacLeod's abandoned house, Ebenezer.

Oriel window detail, Northam.

Front doors, MacNeil House, Tyne Valley. *(see p. 58)*

The MacDonald-Shaw House in Desable
was abandoned for 50 years and is now
demolished.   - PAPEI

Glossary

architrave: the lower division of an entablature, the part that rests directly on the column.

balustrade: a railing composed of posts (balusters) and a handrail.

bargeboard (or vergeboard): boards placed against the eaves of a gable which hide the ends of the horizontal roof members.

belvedere: a small lookout tower on the roof of a house.

bellcast: a roof or eave with a flared edge, shaped like a bell.

board-and-batten: a wall covering of broad vertical boards whose joints are covered by narrow vertical strips of wood.

colonnade: a row of columns.

cornice: an ornamental projecting moulding along the top of a wall or building.

cupola: a small, often domed turret on a circular base, usually set on the ridge of a roof.

dentils: a row of tooth-like blocks, usually in a cornice.

dovetail: an interlocking joint formed by corresponding projections and openings, as in the corners of hewn log buildings.

eave return: the continuation of an eave line in a different direction, usually at a right angle.

eave bracket: a support under a roof beam on the exterior of a house under the eave of a roof or porch.

encaustic tile: earthenware tiles, glazed and decorated.

entablature: the upper part of a Classical order of architecture, comprising architrave, frieze and cornice.

eyebrow dormer: a low, curved dormer on the slope of a roof, sometimes called a "nun's hood" dormer.

fascia: a flat horizontal member or moulding with little projection.

fanlight: a fan-shaped transom window above a door frame, usually found on Classical Revival houses.

filigree: ornamental lacelike work in metal, usually found on the roof ridge or parapet.

finials: an ornament at the top of a roof or decorating the top of an element such as a tower corner.

frieze: a decorative band in a string course, below the cornice.

froe (frow): a wedge-like tool used with a mallet to split timber.

gambrel: a sub-type of the ridge roof with two pitches, the lower one being steeper than the upper.

gingerbread: a decorative woodwork applied to Victorian houses and usually turned on a lathe or cut with a scroll saw.

hood moulding: the projecting moulding of the arch over a door or window, also called a dripstone.

label: a square-arched dripstone or hood mould.

lattice: a structure of crossed wooden or metal strips with open spaces between them.

lintel: a load-bearing horizontal member over an opening such as a window or door.

mansard: a four-sided roof which has a double slope, the lower much steeper than the upper.

mortise-and-tenon: a joint made by fitting a piece of wood with a projection (tenon) into a slot (mortise) in another piece of wood.

newel post: a tall ornamental post at the head or foot of a stair, supporting the handrail.

ogee: arch bounded by reversing curves.

Palladian (Venetian) window: a window in the form of an arch with two additional, narrow, flat-headed side compartments.

pavilion: a prominent portion on a facade, usually central or terminal, identified by projection, height and special roof forms.

pediment: the broad triangular end of a gable, or a triangular element resembling it, on the front of a building.

pilaster: a decorative column applied to a wall.

porte cochère: a doorway large enough to let a vehicle pass; a carriage porch.

portico: a colonnaded porch or roofed area at the entrance of a building.

quoins: cornerstones or blocks forming an outside angle at the junction of two walls.

saddle notch: notches on each timber interlock with corresponding notches in the timbers at right angles to it.

saltbox: a wood-framed house with a short roof pitch in front and a long roof pitch, sweeping close to the ground, in the back.

sidelights: a vertical line of lights or glass panes flanking a doorway.

shelf head: a built-up entablature type head for a window or door.

string course: a projecting horizontal band of decorated bricks on the face of a building.

tracery: ornamental intersecting work in the head or top of a window.

trellis (treillage): an open grating of zigzag latticework made of either metal or wood.

umbrage: a shadowed place, a deep or recessed porch.

vermiculated coursing: a masonry surface, incised with wandering, discontinuous grooves resembling worm tracks.

vernacular architecture: a mode of building based on regional forms and materials.

voussoirs: truncated, wedge-shaped blocks forming an arch.

wainscoting: a decorative or protective facing applied to the lower portion of an interior partition or wall.

wattle and daub: a course basketwork of twigs woven between upright poles, then plastered with mud. Also known as "entorchis".

Bibliography

Adamson, Anthony, and Marion MacRae. *The Ancestral Roof*. Toronto: Irwin, 1963.

Adamson, Anthony, and John Willard. *The Gaiety of Gables*. Toronto: McLelland and Stewart, 1980.

Blanchard, J. Henri. *The Acadians of Prince Edward Island, 1720-1964*. Charlottetown: 1964, 1976.

Blumenson, John J.G. *Identifying American Architecture*. Nashville: American Association for State and Local History, 1977.

Brosseau, Mathilde. *Gothic Revival in Canadian Architecture*. Canadian Historic Sites, Occasional Papers, #25, Ottawa: Parks Canada, 1980.

Brunskill, R.W. *Illustrated Handbook of Vernacular Architecture*. London: Faber, 1971.

Cameron, Christina, and Janet Wright. *Second Empire Style in Canadian Architecture*. Canadian Historic Sites, Occasional Papers, #24. Ottawa: Parks Canada, 1980.

Canadian Heritage Magazine. Volume 14, Issue 2. Summer 1988.

Clark, A.H. *Three Centuries and the Island*. Toronto: University of Toronto Press, 1959.

Clayton, Lucy. *Architecture of Early Prince Edward Island Farmhouses*. (Privately printed.)

Clerk, Nathalie. *Palladian Style in Canadian Architecture*. Studies in Archaeology, Architecture and History. Ottawa: Parks Canada, 1984.

Connally, Ernest Allen. *The Cape Cod House: An Introductory Study*. Urbana, Illinois: University of Illinois, [1960]?

Cullen, Mary. *Pre-1755 Acadian Building Techniques*. Parks Canada Agenda Paper.

Cunningham, Robert, and John B. Prince. *Tamped Clay and Saltmarsh Hay*. Fredericton: Brunswick, 1976.

Daigle, Jean. *Les Acadiens des Maritimes*. Moncton: Centre d'Études Acadiennes, 1980.

Denys, Nicholas. *Concerning the Ways of the Indians*. Halifax: Nova Scotia Department of Education, 1972. (Reprinted.)

Dictionary of Canadian Biography, Volumes 5 and 8. Toronto: University of Toronto Press, 1966.

Downing, A.J. *The Architecture of Country Houses*. New York: Appleton, 1850. Dover, 1969.

Dupont, J.-Claude. *Histoire Populaire de L'Acadie*. Montreal: Lemeac, 1978.

Ennals, Peter. *The Yankee Origins of Bluenose Vernacular Architecture*. American Revue of Canadian Studies, XII-2, Summer 1982.

Ennals, Peter. *The Folk Legacy in Acadian Domestic Architecture: A Study in Mislaid Self Images*. Dimensions of Canadian Architecture. Volume 6. 1983.

Ennals, Peter, and Deryck Holdsworth. *Vernacular Architecture and the Cultural Landscape of the Maritime Provinces - A Reconnaissance*. Acadiensis, Spring/Summer, 1981. pp. 86-106.

Glassie, Henry. *Patterns in the Material Folk Culture of the Eastern U.S.* Philadelphia: University of Pennsylvania Press, 1969.

Gowans, Alan. *Building Canada*. Toronto: Oxford, 1966.

Gowans, Alan. "New England Architecture in Nova Scotia." *The Art Quarterly.* Spring 1962.

Hamilton, William B. *Local History in Atlantic Canada*. pp. 161-179. Toronto: MacMillan, 1974.

Harris, Cyril M. *Historic Architecture Sourcebook*. New York: McGraw Hill, 1977.

Humphreys, Barbara A., and Meredith Sykes. *The Buildings of Canada*. Ottawa: Parks Canada, 1974.

The Island Magazine:
Number 1. Rogers, Irene. *Island Homes*.
  pp. 9-13. Fall/Winter 1976.
Number 4. Tuck, R.C. *Georgetown*.
  pp. 22-28. Spring/Summer 1978.
Number 5. Tuck, R.C. *Bedeque*.
  pp. 15-21. Fall/Winter 1978.
Number 6. Tuck, R.C. *Tignish*.
  pp. 21-26. Spring/Summer 1979.
Number 7. Tuck, R.C. *Victoria*.
  pp. 38-44. Fall/Winter 1979.
Number 8. Tuck, R.C. *Guernsey Cove*.
  pp. 15-20. 1980.
Number 9. Ledwell, Frank J. *Dr. Roddie*.
  pp. 9-14. Fall/Winter 1984.
Number 16. Tuck, R.C. *Seeing Souris*.
  pp. 9-14. Fall/Winter 1984.
Number 18. Morrow, Marianna. *The Builder, Isaac Smith*. pp. 17-23. Fall/Winter 1985.
Number 20. Morrison, J. Clinton Jr. *D.R. Morrison*. pp. 13-18. Fall/Winter 1986.

Maitland, Leslie. *Neoclassical Architecture in Canada*. Studies in Archaeology, Architecture and History. Ottawa: Parks Canada, 1984.

McArdle, Alma, and Deirdre McArdle. *Carpenter Gothic*. New York: Whitney, 1978.

McGee, Harold, and Ruth Holmes Whitehead. *The Micmac*. Halifax: Nimbus, 1983.

Meachem, J.H. and Co. *Illustrated Historical Atlas of Prince Edward Island*. Charlottetown: 1880. Centennial Edition, 1973.

Moogk, Peter. *Building a House in New France.* Toronto: McLelland and Stewart, 1977.

Pillsbury, Richard. *A Field Guide to the Architecture of the Northeastern U.S.* Geography Publications at Dartmouth, Number 8.

Profit, Gloria Jean. "Interesting Architecture and History Found in Some Summerside Homes." *Journal-Pioneer.* October 27, 1980.

Rogers, Irene L. *Charlottetown - The Life in its Buildings.* Charlottetown: Prince Edward Island Museum and Heritage Foundation, 1983.

Rogers, Irene L. *Heritage in Building.* The Canadian Collector; Prince Edward Island Centenary Issue, Volume 8, 1973. pp. 26-30.

Rogers, Irene L. "Historic Island Architecture." A series of weekly articles in the *Charlottetown Guardian* from January 26, 1976, to July 26, 1976.

Rogers, Irene L. *Reports on Selected Buildings in Charlottetown, Prince Edward Island.* Manuscript Report Number 269, National Historic Sites. Ottawa: Parks Canada, 1974, 1976.

Scully, Vincent J. Jr. *The Shingle Style and the Stick Style.* New Haven and London: Yale University Press, 1955, 1971.

Sharpe, Errol. *A People's History of Prince Edward Island.* Toronto: Steel Rail, 1976.

Smith, H.M. Scott. "The Sandstone Houses of Prince Edward Island." *Canadian Antiques and Art Review.* Postal Strike Edition, September 1981. pp. 56-59, 80.

Tuck, Robert C. *Gothic Dreams.* Toronto: Dundurn, 1978.

Warburton, A.B. *A History of Prince Edward Island.* Saint John, N.B.: Barnes, 1923.

Whiffen, Marcus. *American Architecture Since 1780 - A Guide to the Styles.* Cambridge, Massachusetts: M.I.T. Press, 1969.

Wright, Janet. *Architecture of the Picturesque in Canada.* Studies in Archaeology, Architecture and History. Ottawa: Parks Canada, 1984.

Wood, Kathy. "The Witter-Coombs House in Charlottetown." *Canadian Antiques and Art Review.* February 1980. pp. 30-32.

The author wishes to acknowledge references to community histories too numerous to list here and to articles published in the following newspapers: *The Examiner*, the *Charlottetown Guardian-Patriot*, *The Islander*, the *Charlottetown Herald*, the *Prince Edward Island Register* and the *Journal Pioneer*. The author has specific information on these references, if required, and welcomes correspondence.

INDEX

– Chris Reardon

About the Author

Scott Smith is an architect and journalist with a practice in Halifax, Nova Scotia. Born in Montreal, Quebec, he received a Bachelor of Science degree from Mount Allison University in 1967, and a Bachelor of Architecture degree from the School of Architecture, Technical University of Nova Scotia, in 1972. He lived and worked in Charlottetown from 1978 to 1981, during which period he conducted research for this, his second book.

An active conservationist, he has developed a system for recording and evaluating historic buildings. He has lectured on Prince Edward Island architecture at the 1980 Atlantic Canada Institute Conference at the University of Prince Edward Island, conducted a photo survey of historic buildings in Colchester County, Nova Scotia, and his photographs have appeared in many architectural publications. His articles on architectural history have been published by such periodicals as *Arts Atlantic* (Spring 1985) and *Canadian Antiques and Art Review* (September 1981), and he has written book reviews for *The Island* Magazine and *The Atlantic Provinces Book Review*.

SSP
PUBLICATIONS
Samson Smith